T0189436

Lecture Notes in Computer Science　14470

Founding Editors

Gerhard Goos
Juris Hartmanis

Editorial Board Members

The series Lecture Notes in Computer Science (LNCS), including its subseries Lecture Notes in Artificial Intelligence (LNAI) and Lecture Notes in Bioinformatics (LNBI), has established itself as a medium for the publication of new developments in computer science and information technology research, teaching, and education.

LNCS enjoys close cooperation with the computer science R & D community, the series counts many renowned academics among its volume editors and paper authors, and collaborates with prestigious societies. Its mission is to serve this international community by providing an invaluable service, mainly focused on the publication of conference and workshop proceedings and postproceedings. LNCS commenced publication in 1973.

Verónica Vasconcelos · Inês Domingues ·
Simão Paredes
Editors

Progress in Pattern Recognition, Image Analysis, Computer Vision, and Applications

26th Iberoamerican Congress, CIARP 2023
Coimbra, Portugal, November 27–30, 2023
Proceedings, Part II

 Springer

Editors
Verónica Vasconcelos
Polytechnic Institute of Coimbra, Coimbra
Institute of Engineering
Coimbra, Portugal

Inês Domingues 🆔
Polytechnic Institute of Coimbra, Coimbra
Institute of Engineering
Coimbra, Portugal

Simão Paredes 🆔
Polytechnic Institute of Coimbra, Coimbra
Institute of Engineering
Coimbra, Portugal

ISSN 0302-9743   ISSN 1611-3349 (electronic)
Lecture Notes in Computer Science
ISBN 978-3-031-49248-8   ISBN 978-3-031-49249-5 (eBook)
https://doi.org/10.1007/978-3-031-49249-5

Preface

The 26th Iberoamerican Congress on Pattern Recognition (CIARP) was the 2023 edition of the annual international conference CIARP, which aims at fostering international collaboration and knowledge exchange in the fields of pattern recognition, artificial intelligence, and related areas with contributions covering a broad spectrum of theory and applications. We are pleased to acknowledge the endorsement of CIARP 2023 by IAPR, the International Association for Pattern Recognition.

Over the years, CIARP has evolved into a pivotal research event, playing a vital role within the Iberoamerican pattern recognition community. As in previous editions, CIARP 2023 brings together researchers and experts from around the world to showcase ongoing research in areas such as Biometrics, Character recognition, Classification clustering ensembles and multi-classifiers, Data mining and big data, Feature extraction, discretization and selection, Fuzzy logic and fuzzy image processing, Gesture recognition, Hybrid methods, Image description and registration, Image enhancement, restoration and segmentation, Image understanding, Image fusion, Information theory, Intelligent systems, Machine vision, Neural network architectures, Object recognition, Pattern recognition applications, Sensors and sensor fusion, Soft computing techniques, Statistical methods, Syntactical methods, Deep learning, Transfer learning, and Natural language processing.

Moreover, CIARP 2023 serves as a platform for the global scientific community to share their research experiences, disseminate novel insights, and foster collaborations among research groups specialising in artificial intelligence, pattern recognition, and related fields.

CIARP has always prided itself on its international character, and this edition received contributions from 21 countries. Among the Iberoamerican contributors were Portugal, Brazil, Spain, Argentina, Chile, Cuba, Ecuador, Mexico, and Uruguay. Other notable submissions came from France, Germany, Ireland, Belgium, India, South Korea, the Netherlands, Czech Republic, Italy, Taiwan, Tunisia, and the USA.

Through a meticulous review process, involving 59 dedicated reviewers who invested significant time and effort, 61 papers were selected for inclusion in these proceedings, reflecting an acceptance rate of 60.4%. All accepted papers achieved scientific quality scores exceeding the overall mean rating. The selection of reviewers was guided by their expertise, ensuring representation from diverse countries and institutions worldwide. We extend our heartfelt gratitude to all members of the Program Committee for their invaluable contributions, which undoubtedly enhanced the quality of the selected papers.

The conference, held in Coimbra Institute of Engineering, Portugal, from November 27 to 30, 2023, featured four days of engaging sessions, tutorials, and keynotes. The keynotes were delivered by João Paulo Papa, Petia Radeva, and João Manuel R. S. Tavares. CIARP 2023 also awarded the Aurora Pons-Porrata Medal, honouring a female researcher for her significant contributions in the field of pattern recognition and related areas. The authors of the Best Paper and the Best Student Paper and the recipient of the

Aurora Pons-Porrata Medal were invited to submit a paper for publication in the Pattern Recognition Letters journal.

CIARP 2023 was jointly organized by the Coimbra Institute of Engineering (ISEC) and the Polytechnic University of Coimbra (IPC). We express our sincere gratitude for their invaluable contributions to the success of CIARP 2023. We would also like to express our gratitude to i2A for their generous sponsorship. Furthermore, we wish to acknowledge the dedication of all members of the Organizing and Local Committees for their dedication in orchestrating an outstanding conference and proceedings.

We extend our special thanks to the LNCS team at Springer for their invaluable support and guidance throughout the preparation of this volume.

Finally, our deepest gratitude goes out to all authors who submitted their work to CIARP 2023, including those whose papers could not be accommodated. We trust that these proceedings will serve as a valuable reference for the global pattern recognition research community.

November 2023

Inês Domingues
Verónica Vasconcelos
Simão Paredes

Organization

Conference Chairs

Inês Domingues — Polytechnic Institute of Coimbra, Coimbra Institute of Engineering, Portugal

Verónica Vasconcelos — Polytechnic Institute of Coimbra, Coimbra Institute of Engineering, Portugal

Program Chair

Simão Paredes — Polytechnic Institute of Coimbra, Coimbra Institute of Engineering, Portugal

Local Committee

Cristiana Areias — Polytechnic Institute of Coimbra, Coimbra Institute of Engineering, Portugal

Cristina Caridade — Polytechnic Institute of Coimbra, Coimbra Institute of Engineering, Portugal

Fernando Lopes — Polytechnic Institute of Coimbra, Coimbra Institute of Engineering, Portugal

Frederico Santos — Polytechnic Institute of Coimbra, Coimbra Institute of Engineering, Portugal

Luís Santos — Polytechnic Institute of Coimbra, Coimbra Institute of Engineering, Portugal

Nuno Lavado — Polytechnic Institute of Coimbra, Coimbra Institute of Engineering, Portugal

Nuno Martins — Polytechnic Institute of Coimbra, Coimbra Institute of Engineering, Portugal

Teresa Rocha — Polytechnic Institute of Coimbra, Coimbra Institute of Engineering, Portugal

Technical Support

António Godinho — Polytechnic Institute of Coimbra, Coimbra Institute of Engineering, Portugal

Aurora Pons-Porrata Award Committee

Elisabetta Fersini	Università degli Studi di Milano-Bicocca, Italy
Gabriella Pasi	Università degli Studi di Milano-Bicocca, Italy
Maria Matilde García Lorenzo	Universidad Central "Marta Abreu", Cuba

Program Committee

Paulo Ambrósio	Universidade Estadual de Santa Cruz, Brazil
Cristiana Areias	Polytechnic Institute of Coimbra, Coimbra Institute of Engineering, Portugal
Amadeo José Argüelles	Instituto Politécnico Nacional, Mexico
Joel Arrais	University of Coimbra, Portugal
Felipe de Castro Belém	Unicamp, Brazil
Bárbara Caroline Benato	Unicamp, Brazil
Rafael Berlanga	Universitat Jaume I, Spain
Mara Franklin Bonates	Universidade Federal do Ceará, Brazil
Susana Brás	IEETA, UA, Portugal
Alceu de Souza Britto	Pontifícia Universidade Católica do Paraná, Brazil
Maria Elena Buemi	Universidad de Buenos Aires, Argentina
Pedro Henrique Bugatti	Federal University of São Carlos, Brazil
Pablo Cancela	Udelar, Uruguay
Jaime dos Santos Cardoso	FEUP, Portugal
Cristina Caridade	Polytechnic Institute of Coimbra, Coimbra Institute of Engineering, Portugal
Jesús Ariel Carrasco-Ochoa	INAOE, Mexico
Pedro Couto	University of Trás-os-Montes e Alto Douro, Portugal
António Cunha	Universidade de Trás-os-Montes e Alto Douro, Portugal
Matthew Davies	SiriusXM/Pandora, USA
Inês Domingues	Polytechnic Institute of Coimbra, Coimbra Institute of Engineering, Portugal
Jacques Facon	UFES, Brazil
Alicia Fernández	Universidad de la República, Uruguay
Gustavo Fernandez Dominguez	AIT Austrian Institute of Technology, Austria
Vítor Manuel Filipe	University of Trás-os-Montes e Alto Douro, Portugal
Luis Gomez	Universidad de Las Palmas de Gran Canaria, Spain

Pilar Gómez-Gil   National Institute of Astrophysics, Optics and
                                   Electronics, Mexico
Lio Fidalgo Gonçalves   University of Trás-os-Montes e Alto Douro,
                                   Portugal
Teresa Gonçalves   University of Évora, Portugal
Sónia Gouveia   University of Aveiro, Portugal
Michal Haindl   Institute of Information Theory and Automation,
                                   Czech Republic
Xiaoyi Jiang   University of Münster, Germany
Martin Kampel   TU Wien, Austria
Sang-Woon Kim   Myongji University, South Korea
Vitaly Kober   CICESE, Mexico
Nuno Lavado   Polytechnic Institute of Coimbra, Coimbra
                                   Institute of Engineering, Portugal
Fabricio Lopes   Universidade Tecnológica Federal do Paraná,
                                   Brazil
Fernando Lopes   Polytechnic Institute of Coimbra, Coimbra
                                   Institute of Engineering, Portugal
Alexei Machado   Pontifical Catholic University of Minas Gerais,
                                   Brazil
Luís Marques   Polytechnic Institute of Coimbra, Coimbra
                                   Institute of Engineering, Portugal
Nuno Martins   Polytechnic Institute of Coimbra, Coimbra
                                   Institute of Engineering, Portugal
Alessandro Bof Oliveira   UNIPAMPA, Brazil
Hélder Oliveira   INESC TEC/University of Porto, Portugal
João Paulo Papa   São Paulo State University, Brazil
Simão Paredes   Polytechnic Institute of Coimbra, Coimbra
                                   Institute of Engineering, Portugal
Armando J. Pinho   University of Aveiro, Portugal
Pedro Real   Universidad de Sevilla, Spain
Bernardete Ribeiro   University of Coimbra, Portugal
Teresa Rocha   Polytechnic Institute of Coimbra, Coimbra
                                   Institute of Engineering, Portugal
Mateus Roder   São Paulo State University, Brazil
Priscila Saito   Federal University of São Carlos, Brazil
Frederico Santos   Polytechnic Institute of Coimbra, Coimbra
                                   Institute of Engineering, Portugal
Jefersson Alex dos Santos   University of Sheffield, UK
Rafael Santos   INPE, Brazil
Ana Sequeira   INESC TEC, Portugal
Catarina Silva   University of Coimbra, Portugal
José Serra Silva   CINAMIL, Portugal

Samuel Silva	University of Aveiro, Portugal
Luis Enrique Sucar	INAOE, Mexico
Alberto Taboada-Crispi	UCLV, Cuba
César Teixeira	University of Coimbra, Portugal
Luis Filipe Teixeira	Faculdade de Engenharia da Universidade do Porto, Portugal
Murilo Varges da Silva	IFSP, Brazil
Verónica Vasconcelos	Polytechnic Institute of Coimbra, Coimbra Institute of Engineering, Portugal

Contents – Part II

Contents – Part I

Assessing the Generalizability of Deep Neural Networks-Based Models for Black Skin Lesions

Luana Barros, Levy Chaves, and Sandra Avila$^{(\boxtimes)}$

Recod.ai Lab, Institute of Computing, University of Campinas, Campinas, Brazil
`sandra@ic.unicamp.br`

Abstract. Melanoma is the most severe type of skin cancer due to its ability to cause metastasis. It is more common in black people, often affecting acral regions: palms, soles, and nails. Deep neural networks have shown tremendous potential for improving clinical care and skin cancer diagnosis. Nevertheless, prevailing studies predominantly rely on datasets of white skin tones, neglecting to report diagnostic outcomes for diverse patient skin tones. In this work, we evaluate supervised and self-supervised models in skin lesion images extracted from acral regions commonly observed in black individuals. Also, we carefully curate a dataset containing skin lesions in acral regions and assess the datasets concerning the Fitzpatrick scale to verify performance on black skin. Our results expose the poor generalizability of these models, revealing their favorable performance for lesions on white skin. Neglecting to create diverse datasets, which necessitates the development of specialized models, is unacceptable. Deep neural networks have great potential to improve diagnosis, particularly for populations with limited access to dermatology. However, including black skin lesions is necessary to ensure these populations can access the benefits of inclusive technology.

Keywords: Self-supervision · Skin cancer · Black skin · Image classification · Out-of-distribution

1 Introduction

Skin cancer is the most common type, with melanoma being the most aggressive and responsible for 60% of skin cancer deaths. Early diagnosis is crucial to improve patient survival rates. People of color have a lower risk of developing melanoma than those with lighter skin tones [1]. However, melanin does not entirely protect individuals from developing skin cancer. In fact, acral melanoma, or acrolentiginous melanoma, is the rarest and most aggressive type and occurs more frequently in people with darker skin [2]. This subtype is not related to sun exposure, as it tends to develop in areas with low sun exposure, such as the soles, palms, and nails [3].

When melanoma occurs in individuals with darker skin tones, it is often diagnosed later, making it more challenging to treat and associated with a high

© Springer Nature Switzerland AG 2024
V. Vasconcelos et al. (Eds.): CIARP 2023, LNCS 14470, pp. 1–14, 2024.
https://doi.org/10.1007/978-3-031-49249-5_1

mortality rate. This can be partly explained by the fact that acral areas, especially the feet, are often neglected by dermatologists in physical evaluations because they are not exposed to the sun, leading to misdiagnoses [4]. Therefore, it is common for melanoma to be confused by patients with fungal infections, injuries, or other benign conditions [3]. This is related to the lack of representation of cases of black skin in medical education. Most textbooks do not include images of skin diseases as they appear in black people, or when they do, the number is no more than 10% [5]. This absence can lead to a racial bias in the evaluation of lesions by dermatologists since the same lesion may have different characteristics depending on the patient's skin color[1], significantly affecting the diagnosis and treatment of these lesions [5].

Deep neural networks (DNNs) have revolutionized skin lesion analysis by automatically extracting visual patterns for lesion classification and segmentation tasks. However, training DNNs requires a substantial amount of annotated data, posing challenges in the medical field due to the cost and complexity of data collection and annotation. Transfer learning has emerged as a popular alternative. It involves pre-training a neural network, the encoder, on a large unrelated dataset to establish a powerful pattern extractor. The encoder is fine-tuned using a smaller dataset specific to the target task, enabling it to adapt to skin lesion analysis.

Despite the advantages of transfer learning, there is a risk that the pre-trained representations may not fully adapt to the target dataset [6]. Self-supervised learning (SSL) has emerged as a promising solution. In SSL, the encoder is trained in a self-supervised manner on unlabeled data using pretext tasks with synthetic labels. The pretext task is only used to stimulate the network to create transformations in the images and learn the best (latent) representations in the feature space that describe them. This way, we have a powerful feature extractor network that can be used in some other target task of interest, i.e., downstream task. Furthermore, applying SSL models for diagnosing skin lesions has proven advantageous, especially in scenarios with scarce training data [7].

However, deep learning models encounter challenges related to generalization. The effectiveness of machine learning models heavily relies on the quality and quantity of training data available. Unfortunately, in the current medical landscape, skin lesion datasets often suffer from a lack of diversity, predominantly comprising samples from individuals with white skin or lacking explicit labels indicating skin color. This presents a significant challenge as it can lead to models demonstrating racial biases, performing better in diagnosing lesions that are well-represented in the training data from white individuals while potentially encountering difficulties in accurately diagnosing lesions on black skin.

Evaluating skin cancer diagnosis models on black skin lesions is one step towards ensuring inclusivity and accuracy across diverse populations [8]. Most available datasets suffer from insufficient information regarding skin tones, such as the Fitzpatrick scale — a classification of skin types from 1 to 6 based on a person's ability to tan and their sensitivity and redness when exposed to the sun [9] (Figs. 1 and 2). Consequently, we had to explore alternative approaches to address this issue, leading us to conduct the evaluation based on both skin

---

[1] If you have skin, you can get skin cancer.

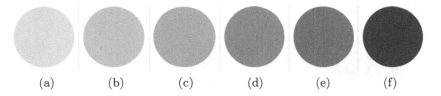

(a)   (b)   (c)   (d)   (e)   (f)

Fig. 1. The Fitzpatrick skin type scale. (a) Type 1 (light): pale skin, always burns, and never tans; (b) Type 2 (white): fair, usually burns, tans with difficulty; (c) Type 3 (medium): white to olive, sometimes mild burn, gradually tans to olive; Type 4 (olive): moderate brown, rarely burns, tans with ease to moderate brown; Type 5 (brown): dark brown, very rarely burns, tans very easily; Type 6 (black): very dark brown to black, never burns, tans very easily, deeply pigmented.

tone and lesion location. We performed two distinct analyses: one focused on directly assessing the impact of skin color using the Fitzpatrick scale, and another centered around evaluating lesions in acral regions, which are more commonly found in individuals with black skin [10].

The primary objective of this work is to assess the performance of skin cancer classification models, which have performed well in white individuals, specifically on black skin lesions. Our contribution is threefold:

- We carefully curate a dataset comprising clinical and dermoscopic images of skin lesions in acral areas (e.g., palms, soles, and nails).
- We evaluate deep neural network models previously trained in a self-supervised and supervised manner to diagnose melanoma and benign lesions regarding two types of analysis:
  - Analysis #1 – Skin Lesions on Acral Regions: We select images from existing datasets focusing on acral regions.
  - Analysis #2 – Skin Lesions in People of Color: We evaluate datasets that contain Fitzpatrick skin type information.
- We have made the curated sets of data and source code available at https:// github.com/httplups/black-acral-skin-lesion-detection.

2 Related Work

The accurate diagnosis of skin lesions in people of color, particularly those with dark skin, has been a long-standing challenge in dermatology. One major contributing factor to this issue is the underrepresentation of dark skin images in skin lesion databases. Consequently, conventional diagnostic tools may exhibit reduced accuracy when applied to this specific population, leading to disparities in healthcare outcomes.

We present a pioneering effort to extensively curate and evaluate the performance of supervised and self-supervised pre-trained models, specifically on black skin lesions and acral regions. While skin lesion classification on acral regions has been explored in previous literature, the focus is largely on general skin types,

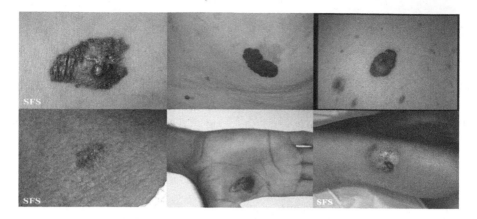

Fig. 2. Each image corresponds to a melanoma sample and is associated with a specific Fitzpatrick scale value, representing a range of skin tones. The images are organized from left to right, following the Fitzpatrick scale (1 to 6). Images retrieved from Fitzpatrick 17k dataset [11].

with limited attention given to black skin tones. Works such as [12–14] investigated classification performance on acral regions, but they do not specifically address the challenges posed by black skin tones.

Addressing the crucial issue of skin type diversity, Alipour et al. [15] conducted a comprehensive review of publicly available skin lesion datasets and their metadata. They observed that only PAD-UFES-20 [16], DDI [17], and Fitzpatrick 17k [11] datasets provide the Fitzpatrick scale as metadata, highlighting the need for improved representation of diverse skin types in skin lesion datasets. However, the authors did not conduct model evaluations on these datasets.

Existing works explored the application of the Fitzpatrick scale in various areas, such as debiasing [18,19] and image generation [20]. However, these studies have not adequately addressed the specific challenge of skin lesion classification on black skin tones.

To bridge this research gap, our study evaluates the performance of supervised and self-supervised pre-trained models exclusively on black skin lesion images and acral regions. By systematically exploring and benchmarking different pre-training models, we aim to contribute valuable insights and advancements to the field of dermatology, particularly in the context of underrepresented skin types.

3 Materials and Methods

In this work, we assess the performance of six pre-trained models on white skin in black skin. We pre-train all models as described in Chaves et al. [7]. First, we take a pre-trained model backbone on ImageNet [21] and fine-tune it on the ISIC dataset [22]. The ISIC (*International Skin Imaging Collaboration*) is a common choice in this domain [6,11,23], presenting only white skin images. Next, we evaluate the fine-tuned model on several **out-of-distribution datasets**, where the distribution of the test data diverges from the training one. We also use the same six pre-trained models as Chaves et al. [7] because they have the code and checkpoint publicity available to reproduce their results. The authors compared the performance of five self-supervised models against a supervised baseline and showed that self-supervised pre-training outperformed traditional transfer learning techniques using the ImageNet dataset.

We use the ResNet-50 [24] network as the feature extractor backbone. The self-supervised approaches vary mainly in the choice of pretext tasks, which are BYOL (*Bootstrap Your Own Latent*) [25], InfoMin [26], MoCo (*Momentum Contrast*) [27], SimCLR (*Simple Framework for Contrastive Learning of Visual Representations*) [28], and SwAV (*Swapping Assignments Between Views*) [29]. We assessed all six models using two different analyses on compound datasets. The first analysis focused on skin lesions in acral regions, while the second considered variations in skin tone. Next, we detail the datasets we curated.

3.1   Datasets

Analysis #1: Skin Lesions on Acral Regions. To create a compound dataset of acral skin lesions, we extensively searched for datasets and dermatological atlases available on the Internet that provided annotations indicating the location of the lesions. We analyzed 17 datasets listed in SkinIA's website[2] then filtered the datasets to include only images showcasing lesions in acral regions, such as the palms, soles, and nails. As a result, we identified three widely recognized datasets in the literature, namely the International Skin Imaging Collaboration (ISIC Archive) [22], the 7-Point Checklist Dermatology Dataset (Derm7pt) [30], and the PAD-UFES-20 dataset [16]. We also included three dermatological atlases: Dermatology Atlas (DermAtlas) [31], DermIS [32], and DermNet [33].

We describe the steps followed for each dataset in the following. Table 1 shows the number of lesions for each dataset.

ISIC Archive [22]: We filtered images from the ISIC Archive based on clinical attributes, focusing on lesions on palms and soles, resulting in 773 images. We excluded images classified as carcinoma or unknown, reducing the dataset to 400. As we trained our models using ISIC Archive, we removed all images appearing in the models' training set to avoid data leakage between training and testing data and ensure an unbiased evaluation, resulting in a final dataset with 149 images.

Derm7pt [30]: It consists of 1011 images for each lesion, including clinical and dermoscopic versions[3]. It offers valuable metadata such as visual patterns,

---

[2] https://www.medicalimageanalysis.com/data/skinia.

[3] Clinical images can be captured with standard cameras, while dermoscopic images are captured with a device called dermatoscope, that normalize the light influence on the lesion, allowing to capture deeper details.

lesion location, patient sex, difficulty level, and 7-point rule scores [34]. We applied a filter based on lesion location to select images from it, selecting acral images from the region attribute. This filter resulted in a total of 62 images, comprising only benign and melanoma lesions. We conducted separate evaluations using the clinical and dermoscopic images, labeling the datasets as *derm7pt-clinic* and *derm7pt-derm*, respectively.

PAD-UFES-20 [16]: It comprises 2298 clinical images collected from smartphone patients. It also includes metadata related to the Fitzpatrick scale, providing additional information about skin tone. We focused on the hand and foot region lesions, which yielded 142 images. We also excluded images classified as carcinoma (malignant), resulting in a final set of 98 images.

Atlases (DermAtlas, DermIS, DermNet): The dataset included images obtained from dermatological atlas sources such as DermAtlas [31], DermIS [32], and DermNet [33]. We use specific search terms, such as *hand*, *hands*, *foot*, *feet*, *acral*, *finger*, *nail*, and *nails* to target the lesion location. We conducted a manual selection to identify images meeting the melanoma or benign lesions criteria. This dataset comprised 8 images from DermAtlas (including 1 melanoma), 12 images from DermIS (comprising 10 melanomas), and 34 images from DermNet, all melanomas. Finally, we combined all images in a set referenced as Atlases, containing 54 images.

Table 1. Number of benign and melanoma lesions for acral areas dataset.

Dataset	Number of Lesions		
	Melanoma	Benign	Total
ISIC Archive [22]	72	77	149
Derm7pt [30]	3	59	62
PAD-UFES-20 [16]	2	96	98
Atlases [31–33]	45	9	54

Analysis #2: Skin Lesions in People of Color. We focused on selecting datasets that provided metadata indicating skin tone to analyze skin cancer diagnosis performance for darker-skinned populations. Specifically, datasets containing skin lesions with darker skin tones (Fitzpatrick scales 4, 5, and 6) allow us to evaluate the performance of the models on these populations. For this purpose, we evaluated three datasets: PAD-UFES-20 [16], which was previously included in the initial analysis, as well as Diverse Dermatology Images (DDI) [17], and Fitzpatrick 17k [11].

Table 2 shows the number of lesions for each dataset, considering the Fitzpatrick scale.

Table 2. Number of benign and melanoma lesions grouped by Fitzpatrick scale for skin tone analysis datasets.

Dataset	Fitzpatrick Scale	Number of Lesions		
		Melanoma	Benign	Total
PAD-UFES-20* [16]	1–2	38	246	284
	3–4	14	153	167
	5–6	0	6	6
	Total	52	405	457
DDI [17]	1–2	7	153	160
	3–4	7	153	160
	5–6	7	134	141
	Total	21	440	461
Fitzpatrick 17k [11]	1–2	331	1115	1446
	3–4	168	842	1010
	5–6	47	203	250
	Total	546	2160	2706

PAD-UFES-20*: We filtered images using the Fitzpatrick scale, including lesions from all regions rather than solely acral areas. We specifically selected melanoma cases from the malignant lesions category, excluding basal and squamous cell carcinomas. Also, we excluded images lacking Fitzpatrick scale information. Consequently, the dataset was refined to 457 images, including 52 melanoma cases. Notably, within this dataset, there were only five images with a Fitzpatrick scale of 5 and one image with a Fitzpatrick scale of 6.

Diverse Dermatology Images (DDI) [17]: The primary objective of DDI is to address the lack of diversity in existing datasets by actively incorporating a wide range of skin tones. For that, the dataset was curated by experienced dermatologists who assessed each patient's skin tone based on the Fitzpatrick scale. The initial dataset comprised 656 clinical images, categorized into different Fitzpatrick scale ranges. We filtered to focus on melanoma samples for malignant lesions. As a result, we excluded benign conditions that do not fall under benign skin lesions, such as inflammatory conditions, scars, and hematomas. This process led to a refined dataset of 461 skin lesions, comprising 440 benign lesions and 21 melanomas. Regarding the distribution based on the Fitzpatrick scale, the dataset includes 160 images from scales 1 to 2, 160 images from scales 3 to 4, and 141 images from scales 5 to 6. The DDI dataset represents a notable improvement in diversity compared to previous datasets, but it still exhibits an unbalanced representation of melanoma images across different skin tones.

Fitzpatrick 17k [11]: It comprises 16,577 clinical images, including skin diagnostic labels and skin tone information based on the Fitzpatrick scale. The dataset was compiled by sourcing images from two online open-source dermatology atlases: 12,672 images from DermaAmin [35] and 3,905 images from Atlas Dermatologico [36]. To ensure the analysis specifically targeted benign and melanoma skin lesion conditions, we applied a filter based on the "nine_partition_attribute". This filter allowed us to select images that fell into benign dermal, benign epidermal, benign melanocyte, and malignant melanoma. After removing images with the unknown Fitzpatrick value, the refined dataset consists of 2,706 images, 191 images corresponding to a Fitzpatrick scale of 5 and 59 images corresponding to a Fitzpatrick scale of 6.

3.2   Evaluation Pipeline

Our pipeline to evaluate skin lesion image classification models is divided into two main stages: pre-processing and model inference. Figure 3 shows the pipeline.

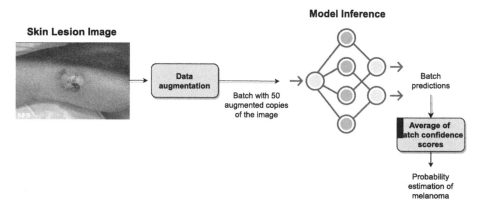

Fig. 3. Evaluation pipeline for all models. Given a test image, we adopt the final confidence score as the average confidence over a batch of 50 augmented copies of the input image.

Pre-Processing: We apply data augmentation techniques to the test data, which have been proven to enhance the performance of classification problems [23]. The test set is evaluated in batches, and a batch of 50 copies is created for each image. Each copy undergoes various data augmentations, including resizing, flipping, rotations, and color changes. Additionally, we normalize the images using the mean and standard deviation values from the ImageNet dataset.

Model Inference: The batch of augmented images is fed into the selected model for evaluation. The model generates representations or features specific to its pre-training method. These representations are then passed through a softmax layer, which produces the probability values for the lesion being melanoma, the positive class of interest. We calculate the average of the probabilities obtained from all 50 augmented copies to obtain a single probability value for each image in the batch.

The evaluation process consisted of two analyses: Analysis #1 (skin lesions on acral regions), which considered acral images, and Analysis #2 (skin lesions in people of color), which considered images with diverse skin tones according to the Fitzpatrick scale. For each analysis, we assessed each dataset individually using the six models: BYOL, InfoMin, MoCo, SimCLR, SwAV, and the Supervised baseline. In each evaluation, the dataset was passed to the respective model, and the probability of melanoma lesions was obtained for all images. Metrics such as balanced accuracy, precision, recall, and F1-score were calculated based on these probabilities. We computed balanced accuracy using a threshold of 0.5.

4 Results

4.1 Skin Lesion Analysis on Acral Regions

Table 3 shows the classification metrics grouped by datasets of SSL models and the supervised baseline for skin lesions in acral regions, such as palms, soles, and nails. In the following, we discussed the results considering each dataset.

Table 3. Evaluation metrics for acral skin lesions. We grouped ISIC Archive, Derm7pt, Atlases, and PAD-UFES-20 due to some datasets' low number of Melanoma samples. #Mel and #Ben indicate the number Melanomas, and benign skin lesions, respectively.

Samples (#Mel/#Ben)	Model	Balanced Accuracy (%)	Precision (%)	Recall (%)	F1-score (%)
(89/336)	SwAV	78.3	**77.9**	64.3	70.4
	MoCo	79.9	72.0	**71.4**	71.7
	SimCLR	76.1	70.2	63.5	66.7
	BYOL	77.9	70.8	67.5	69.1
	InfoMin	**80.2**	74.2	70.6	**72.4**
	Supervised	78.4	73.7	66.7	70.0
	Mean	78.5	73.1	67.3	70.1

ISIC Archive: We observed consistent result between balanced accuracy and F1-score, both averaging around 87%. The evaluation metrics exhibit high performance due to the fine-tuning process of the evaluated models using the ISIC 2019 dataset. The distribution of the ISIC Archive dataset closely resembles that of the training data, distinguishing it from other datasets, and contributing to the favorable evaluation metrics observed, even though excluding training samples from our evaluation set. Furthermore, in the ISIC 2019 dataset, all results were above 90% [7]. This indicates that even with an external dataset with a distribution more akin to the training data, the performance for lesions in acral regions is significantly inferior to that in other regions. Additionally, it is essential to highlight that in the ISIC 2019 dataset, all results exceeded 90% [7].

Derm7pt: We analyzed two types of images: dermoscopic (derm7pt-derm) and clinical (derm7pt-clinical). When examining the F1-score results for clinical images, the models (SwAV, BYOL, and Supervised) encountered challenges in accurately classifying melanoma lesions. However, the evaluation was performed on a limited sample size of only three melanoma images. This scarcity of data for melanoma evaluation has contributed to the observed zero precision and recall scores. On average, dermoscopic images demonstrated better classification performance than clinical images, with dermoscopic images achieving an F1-score of 26% and clinical images achieving an F1-score of 16%. We attribute this disparity to the models being trained on dermoscopic images from the ISIC 2019 dataset. Additionally, using different image capture devices (dermatoscope vs. cell phone camera) can introduce variations in image quality and the level of detail captured, affecting the overall data distribution. Given that the models were trained with dermoscopic images and the test images were captured using a dermatoscope, the training and test data distributions are expected to be more similar. In general, the results for this dataset demonstrated low F1-score and balanced accuracy, indicating an unsatisfactory performance, especially for clinical images.

Atlases: The performance varies across different models. MoCo and InfoMin achieved balanced accuracies of approximately 72%, indicating relatively better performance. Other models, such as Supervised and BYOL, exhibited poor results. Such dataset is considered challenging as it consists of non-standardized skin lesions collected from online atlases, which may introduce variability in the capture process. Still, models could perform better than previous datasets on acral region images, specifically when considering F1-score values.

PAD-UFES-20: The models achieved an average balanced accuracy of around 90%. The model SwAV performed best, with a balanced accuracy of 95.8% and an F1-score of 33.3%. All models showed similar patterns: the F1-score and precision were relatively low, while recall was high (100%). The high recall was mainly due to the correct prediction of the two melanoma samples in the dataset, which inflated the balanced accuracy score. It indicates that relying solely on balanced accuracy can lead to a misleading interpretation of the results. Also, the small number of positive class samples limits the generalizability of the results and reduces confidence in the evaluation.

4.2   Skin Lesion Analysis in People of Color

Table 4 shows the evaluation results of the SSL models and the Supervised baseline for datasets containing melanoma and benign black skin lesions.

DDI: revealed poor results regarding balanced accuracy and F1-score for all models. The supervised baseline model performed the worst, with an F1-score of only 3.4%, while MoCo achieved a slightly higher F1-score of 12.5%. Although most of the DDI dataset consisted of benign lesions, the performance of all

Table 4. Evaluation metrics for skin tone analysis. #Mel and #Ben indicate the number Melanomas, and benign skin lesions, respectively.

Dataset (#Mel/#Ben)	Model	Balanced Accuracy (%)	Precision (%)	Recall (%)	F1-score (%)
DDI (21/440)	SwAV	52.9	7.5	14.3	9.8
	MoCo	**55.8**	**8.5**	**23.8**	**12.5**
	SimCLR	54.2	7.8	19.0	11.1
	BYOL	54.3	8.0	19.0	11.3
	InfoMin	54.4	8.2	19.0	11.4
	Supervised	48.2	2.6	4.8	3.4
	Mean	53.3	7.1	16.7	9.9
Fitzpatrick 17k (546/2160)	SwAV	57.6	40.7	24.2	30.3
	MoCo	59.8	38.4	32.1	34.9
	SimCLR	59.3	40.7	29.5	34.2
	BYOL	59.3	38.4	32.1	34.9
	InfoMin	60.1	36.5	**35.9**	36.2
	Supervised	**63.4**	**51.2**	35.3	**41.8**
	Mean	60.0	40.4	32.4	35.6
PAD-UFES-20* (52/405)	SwAV	57.1	25.0	23.1	24.0
	MoCo	59.1	23.9	30.8	26.9
	SimCLR	58.4	21.0	**32.7**	25.6
	BYOL	**59.2**	**26.3**	28.8	**27.5**
	InfoMin	54.3	16.9	23.1	19.5
	Supervised	58.5	23.8	28.8	26.1
	Mean	57.8	22.8	27.9	24.9

models was considered insufficient. This underscores the significance of a pre-training process incorporating diverse training data, as it enables the models to learn more robust and generalizable representations across different skin tones and lesion types. In addition, this highlights the importance of self-supervised learning in improving performance and diagnostic accuracy, particularly in the context of diverse skin tones.

Fitzpatrick 17k: In contrast to the DDI dataset, the supervised model achieved the highest performance in balanced accuracy (63.4%) and F1-score (41.8%). Both self-supervised and supervised models showed similar results for this dataset.

PAD-UFES-20:* Both self-supervised and supervised models demonstrated comparable performance. The BYOL method achieved the highest balanced accuracy (59.2%) and F1-score (27.5%). It is essential to highlight that this dataset did not include any melanoma lesions corresponding to the Fitzpatrick scale of 5 and 6 (see Table 2).

5   Conclusion

Our evaluation of self-supervised and supervised models on skin lesions in acral regions reveals a significant deficiency in robustness and bias in deep-learning models for out-of-distribution images, especially in darker skin tones. Both Self-supervised and Supervised models achieved poor performance in Melanoma classification task compared to white skin only datasets. These results highlight the generalization gap between models trained on white skin and tested on darker skin tones, inviting further work on improving the generalization capabilities of such models. But, we believe that improvements are not only necessary in model designing, but requires richer data to represent specific population or subgroups.

The results for melanoma diagnosis in acral regions are insufficient and could cause serious social problems if used clinically. Additionally, more samples are needed to improve the metrics calculation and analysis of results. The generalization power of DNNs-based models heavily depends on training data distribution. Therefore, for DNNs-based models to be robust concerning different visual patterns of lesions, training them with datasets that represent the real clinical scenario, including patients with diverse lesion characteristics and skin tones, is necessary. There is an urgent need for the creation of datasets that guarantee data transparency regarding the source, collection process, and labeling of lesions, as well as the reliability of data descriptions and the ethnic and racial diversity of patients, in order to ensure high confidence in the diagnoses made by the models.

The current state of skin cancer datasets is concerning as it impacts the performance of models and can further reinforce biases in diagnosing skin cancer in people of color. Currently, these models cannot be used in a general sense, as they only perform well on lesions in white skin on common regions affected, and their performance may vary significantly for people with different skin tones. Crafting models that are discriminative for diagnoses, yet discriminate against patients' skin tones, is unacceptable. Deep neural networks have great potential to improve diagnosis, especially for populations with limited access to dermatology. However, including black skin lesions is extremely necessary for these populations to access the benefits of inclusive technology.

Acknowledgments. L. Chaves is funded by Becas Santander/Unicamp - HUB 2022, Google LARA 2021, in part by the Coordenação de Aperfeiçoamento de Pessoal de Nível Superior - Brasil (CAPES) - Finance Code 001. S. Avila is funded by CNPq 315231/2020-3, FAEPEX, FAPESP 2013/08293-7, 2020/09838-0, H.IAAC 01245.013778/2020-21, and Google Award for Inclusion Research Program 2022 ("Dark Skin Matters: Fair and Unbiased Skin Lesion Models").

References

1. American Cancer Society. Key statistics for melanoma skin cancer (2022). https://www.cancer.org/cancer/melanoma-skin-cancer/about/key-statistics.html
2. AIM at Melanoma Foundation. What is acral lentiginous melanoma? https://www.aimatmelanoma.org/melanoma-101/types-of-melanoma/cutaneous-melanoma/acral-lentiginous-melanoma/
3. Memorial Sloan Kettering Cancer Center. Types of melanoma (2022). https://www.mskcc.org/cancer-care/types/melanoma/types-melanoma
4. Caetano, Y.A., Quinteiro Ribeiro, A.M., da Silva Albernaz, B.R., de Paula Eleutério, I., Fleury Fróes, L.F.: Melanoma acral-estudo clínico e epidemiológico. Surgical & Cosmetic Dermatology (2020)
5. Rabin, R.C.: Dermatology has a problem with skin color (2020). https://www-nytimes-com.cdn.ampproject.org/c/s/www.nytimes.com/2020/08/30/health/skin-diseases-black-hispanic.amp.html
6. Menegola, A., Fornaciali, M., Pires, R.: Flávia Vasques Bittencourt, Sandra Avila, and Eduardo Valle. Knowledge transfer for melanoma screening with deep learning, International Symposium on Biomedical Imaging (2017)
7. Chaves, L., Bissoto, A., Valle, E., Avila, S.: An evaluation of self-supervised pre-training for skin-lesion analysis. In: European Conference on Computer Vision Workshops (2022)
8. Singh, N.: Decolonising dermatology: why black and brown skin need better treatment. The Guardian, 13 (2020)
9. DermNet. Fitzpatrick skin phototype (2012). https://dermnetnz.org/topics/skin-phototype
10. Dermatology Learning Network. Skin cancer in African-Americans (2004). https://www.hmpgloballearningnetwork.com/site/thederm/article/2547
11. Groh, M., et al.: Evaluating deep neural networks trained on clinical images in dermatology with the fitzpatrick 17k dataset. In: Conference on Computer Vision and Pattern Recognition (2021)
12. Chanki, Yu., et al.: Acral melanoma detection using a convolutional neural network for dermoscopy images. PloS one (2018)
13. Lee, S., et al.: Augmented decision-making for acral lentiginous melanoma detection using deep convolutional neural networks. J. Eur. Acad. Dermatology Venereology (2020)
14. Abbas, Q., Ramzan, F., Ghani, M.U.: Acral melanoma detection using dermoscopic images and convolutional neural networks. Visual Computing for Industry, Biomedicine, and Art (2021)
15. Alipour, N., Burke, T., Courtney, J.: Skin type diversity: a case study in skin lesion datasets (2023)
16. Pacheco, A., et al. PAD-UFES-20: a skin lesion dataset composed of patient data and clinical images collected from smartphones. Data in Brief (2020)
17. Daneshjou, R., et al.: Disparities in dermatology AI performance on a diverse, curated clinical image set. Science Advances (2022)
18. Bevan, P.J., Atapour-Abarghouei, A.: Detecting melanoma fairly: skin tone detection and debiasing for skin lesion classification. In: MICCAI Workshop on Domain Adaptation and Representation Transfer, pp. 1–11 (2022)
19. Pakzad, A., Abhishek, K., Hamarneh, G.: Circle: color invariant representation learning for unbiased classification of skin lesions. In: European Conference on Computer Vision (2022)

20. Rezk, E., Eltorki, M., El-Dakhakhni, W., et al.: Improving skin color diversity in cancer detection: deep learning approach. JMIR Dermatology 5(3), e39143
21. Deng, J., Dong, W., Socher, R., Li, L.-J., Li, K., Fei-Fei, L.: Imagenet: a large-scale hierarchical image database. In: Conference on Computer Vision and Pattern Recognition (2009)
22. ISIC Archive (2023). https://www.isic-archive.com
23. Valle, E., et al.: Data, depth, and design: Learning reliable models for skin lesion analysis. Neurocomputing (2020)
24. He, K., Zhang, X., Ren, S., Sun, J.: Deep residual learning for image recognition. In: IEEE Conference on Computer Vision and Pattern Recognition, pp. 770–778 (2016)
25. Grill, J.-B., et al.: Bootstrap your own latent - a new approach to self-supervised learning. In: Advances in Neural Information Processing Systems (2020)
26. Tian, Y., Sun, C., Poole, B., Krishnan, D., Schmid, V., Isola, P.: What makes for good views for contrastive learning? In: Advances in Neural Information Processing Systems (2020)
27. He, K., Fan, H., Wu, Y., Xie, S., Girshick, R.: Momentum contrast for unsupervised visual representation learning. In: Conference on Computer Vision and Pattern Recognition (2020)
28. Chen, T., Kornblith, S., Norouzi, M., Hinton, G.: A simple framework for contrastive learning of visual representations. In: International Conference on Machine Learning (2020)
29. Caron, M., Misra, I., Mairal, J., Goyal, P., Bojanowski, P., Joulin, A.: Unsupervised learning of visual features by contrasting cluster assignments. Advances in Neural Information Processing Systems (2020)
30. Kawahara, J., Daneshvar, S., Argenziano, G., Hamarneh., H.: Seven-point checklist and skin lesion classification using multitask multimodal neural nets. IEEE J. Biomed. Health Inf. (2019)
31. Richard, P.: Usatine and Brian D. Madden. Interactive dermatology atlas (2023). https://www.dermatlas.net
32. Dermis.net: Dermatology information service available on the internet (2023). https://www.dermis.net/dermisroot/pt/home/index.htm
33. Dermnet resource (2023). https://dermnetnz.org
34. Argenziano, G., Fabbrocini, G., Carli, P., De Giorgi, V., Sammarco, E., Delfino, M.: Epiluminescence microscopy for the diagnosis of doubtful melanocytic skin lesions: comparison of the ABCD rule of dermatoscopy and a new 7-point checklist based on pattern analysis. Arch. Dermatol. 134(12), 1563–1570 (1998)
35. AlKattash, J.A.: Dermaamin. https://www.dermaamin.com
36. da Silva, S.F.: Atlas dermatologico. http://atlasdermatologico.com.br

Breast MRI Multi-tumor Segmentation Using 3D Region Growing

Teresa M. C. Pereira[1,2,4(✉)], Ana Catarina Pelicano[2], Daniela M. Godinho[2], Maria C. T. Gonçalves[2], Tiago Castela[3], Maria Lurdes Orvalho[3], Vitor Sencadas[4], Raquel Sebastião[1,5], and Raquel C. Conceição[2]

[1] Institute of Electronics and Informatics Engineering of Aveiro (IEETA), Department of Electronics, Telecommunications and Informatics (DETI), Intelligent Systems Associate Laboratory (LASI), University of Aveiro, 3810-193 Aveiro, Portugal
teresamcp@ua.pt
[2] Instituto de Biofísica e Engenharia Biomédica, Faculdade de Ciências da Universidade de Lisboa, Campo Grande, 1749-016 Lisbon, Portugal
[3] Departamento de Radiologia, Hospital da Luz Lisboa, Luz Saúde, 1500-650 Lisbon, Portugal
[4] Departamento de Engenharia de Materiais e Cerâmica, Instituto de Materiais (CICECO), Universidade de Aveiro, 3810-193 Aveiro, Portugal
[5] Polytechnic of Viseu, 3504-510 Viseu, Portugal

Abstract. Breast tumor is one of the most prominent indicators for diagnosis of breast cancer. Magnetic Resonance Imaging (MRI) is a relevant imaging modality tool for breast cancer screening. Moreover, an accurate 3D segmentation of breast tumors from MRI scans plays a key role in the analysis of the disease. This paper presents a pipeline to automatically segment multiple tumors in breast MRI scans, following the methodology proposed by one previous study, addressing its limitations in detecting multiple tumors and automatically selecting seed points using a 3D region growing algorithm. The pre-processing includes bias field correction, data normalization, and image filtering. The segmentation process involved several steps, including identifying high-intensity points, followed by identifying high-intensity regions using k-means clustering. Then, the centers of the regions were used as seeds for the 3D region growing algorithm, resulting in a mask with 3D structures. These masks were then analyzed in terms of their volume, compactness, and circularity. Despite the need for further adjustments in the model parameters, the successful segmentation of four tumors proved that our solution is a promising approach for automatic multi-tumor segmentation with the potential to be combined with a classification model relying on the characteristics of the segmented structures.

Keywords: Magnetic Resonance Imaging · Breast Tumor · Tumor Segmentation · Region Growing

© Springer Nature Switzerland AG 2024
V. Vasconcelos et al. (Eds.): CIARP 2023, LNCS 14470, pp. 15–29, 2024.
https://doi.org/10.1007/978-3-031-49249-5_2

1 Introduction

Breast cancer was the most common type of cancer diagnosed around the world in 2020, with over 2.26 million new cases. It was also reported as the cancer with the highest mortality rate in women and the fifth highest for both genders [1]. It is estimated to be responsible for 15% of cancer deaths [3]. Early detection and intervention have been shown to significantly reduce the mortality rate and improve the quality of life and survival rates of breast cancer patients. Therefore, these factors are considered crucial for successful treatment outcomes [9].

In order to assist radiologists in detecting masses in early stages of breast cancer, it is highly desirable to develop a reliable Computer-aided Diagnosis (CAD) system. These systems have become increasingly advanced and are now routinely used in clinical settings [6]. They typically involve three stages: detection, segmentation, and classification of masses. Automatic segmentation of breast tissue in MRI image is a two-step process. First, the breast area is separated from the chest wall and pectoral muscles (outer segmentation). In the second step, the breast tissue is further divided into fibro-glandular, fatty and tumor tissue (inner segmentation). This automatic segmentation of breast tissue can be challenging due to variations in breast size and shape, intensity inhomogeneities, image artifacts, and other noise errors [12]. Tumors are then classified as either malignant or benign based on their characteristics. Malignant tumors are often irregularly shaped and surrounded, by spicules, whereas benign tumors tend to have more rounded or elliptical shapes [10].

In this study, a segmentation pipeline proposed by Pelicano et al. [9] was used to segment highly heterogeneous tumors from MRI exams. However, this study presented two limitations: 1) the need for manual selection of the seed point; and 2) the inability to segment multiple tumors, as the 3D region growing algorithm is only capable of identifying one tumor per seed of high intensity. This limitation is addressed by proposing an alternative segmentation process in which multi-tumor classification is performed using an automatic selection of the seed points for the 3D region growing algorithm. The proposed pipeline includes the following steps: (i) image pre-processing and (ii) tumor segmentation. This paper is organized as follows: firstly, an overview of related work on breast tumor segmentation is presented; then, a description of the materials and methodology used for image pre-processing and tumor segmentation is provided. Then, the results of the proposed methods are presented, followed by a discussion of the findings. Finally, the main conclusions of this work are highlighted.

2 Literature Review

Properly identifying the boundaries of a tumor is crucial to evaluate its characteristics and determine appropriate treatment. However, this task can be quite challenging due to the wide variability in shape and intensity distribution of breast lesions.

Traditional methodologies to identify the region of interest (tumor) in an image typically focus on using the intensity values of the entire image. This is based on the assumption that biological tissues are typically well-separated in grayscale images. However, when dealing with heterogeneous structures that contain a wide range of intensity values, such as diverse malignant tumors, these methods may not be able to group all tumor voxels into the same cluster, resulting in a poor segmentation of the tumor volume. Researchers have proposed various methods to circumvent this limitation, such as by specifying a seed point where the algorithm begins the segmentation process and a threshold value to distinguish lesion and non-lesion regions.

Yin et al. [16] proposed a technique to reconstruct the 3D volume of a breast tumor using segmented 2D slices. The method involved using seeded region growing and Otsu thresholding to eliminate the breast border region on each 2D slice, followed by segmentation using a gradient-based level set approach. The resulting segmented slices were then processed using a ray-casting method to rebuild the 3D volume of the tumor. Chen et al. [4] presented a technique to create a multi-faceted depiction of the contours of extracted breast tumors in three dimensions. Their approach consisted of using a 2D level set method for the segmentation of individual slices, applying this method to three distinct planes - transverse, coronal, and sagittal. Specifically, the method located the targeted tumor primarily in certain slices of an MRI and then used these identified locations to determine the tumor's position in the remaining slices. The 2D contour of the tumor was then highlighted in each slice through the use of the 2D level set method. Finally, these found 2D contours were combined to construct a 3D representation of the tumor contours for the specific plane under examination.

Wang et al. [15] proposed a deep learning model to identify micro-calcifications in mammograms. The model uses a technique of dividing the image into multiple small clusters, each of these clusters is then separately evaluated, where the small clusters are used to focus on the characteristics of the micro-calcifications and the larger clusters are used to analyze the surrounding tissue. In [14], authors also used a machine learning algorithm for the detection of breast tumors in mammograms. The location of the tumor was determined based on a combination of geometric features such as roundness, entropy, ratio of area, variance, and roughness, as well as texture features including energy, inverse difference, correlated coefficient, and contrast. Melouah et al. [7] presented a method to segment mammograms using a combination of threshold and seed point selection. The threshold value was determined by finding the mean of the maximum values in each row of the pixel matrix. For seed point selection, various statistical features such as mean, standard deviation, contrast, entropy, regularity, and uniformity were calculated, and the mean of these features was used as the initial seed point. This method produced good results in terms of detecting tumors, but it struggled in cases where the mammogram contained multiple tumors. Later, Shrivastava et al. [11] presented a technique to identify the specific Region Of Interest (ROI) in mammograms through automatic seed point identification and threshold calculation using the seeded region growing method. Similarly, Al-Faris et al. in [2] studied the segmentation of tumors in MRI images using a modified version of the automatic seeded region growing algorithm, incorporating variations in seed point and threshold selection for improved performance compared to previous methods. More recently, Pelicano et al. [9] proposed a method to segment tumors in MRI images using a 3D version of the region growing technique. This method is similar to seeded region growing but operates in three dimensions. The method is semi-automatic as it needs a manual selection of the seed but it is highly successful in segmenting tissues with different levels of heterogeneity. However, the algorithm fails to detect the presence of multiple tumors.

Although various methods have been proposed in the literature for tumor segmentation, most of them present some limitations. Many of these methods prioritize high segmentation accuracy over computation efficiency, which can lead to high computational costs. Additionally, some approaches require a significant level of human input, such as the selection of initial seed positions and threshold values, making them less practical to use. In contrast, some others achieve high segmentation accuracy but fail to segment multi-tumor.

3   Materials and Methods

In this project, two MRI scans from two patients were used, yielding a total of six tumors. One of the scans had five tumors, while the other had just one. The patients were imaged in a prone position using a 3.05 MAGNETON Vida clinical Magnetic Resonance (MR) scan from Siemens Healthineers, Erlangen, Germany. The scan was performed with the use of a specialized coil for the breast, called the Siemens Breast 18 coil, also from Siemens Healthineers. The images were collected in Hospital de Luz Lisboa under clinical protocols: CES/44/2019/ME (2019) and CES/34/2020/ME (2020).

Two MRI sequences were collected: Dynamic Contrast-Enhanced transversal three-dimensional T1-weighted Fast low Angle Shot 3D Spectral Attenuated Inversion Recovery sequence (DCE-fl3D); and Direct coronal isotropic 3D T1-w fl3D Volumetric Interpolated Breath-hold Examination (VIBE) Dixon image sequence (T1-w Dixon). DCE-fl3D consists of a fat-suppression sequence with six sets of images: a pre-contrast image, acquired before the injection of intravenous contrast agent gadolimium, and five post-contrast consecutive images. The contrast agent enhances highly vascularized tissues, such as tumors, allowing them to stand out in the image. The digital subtractions enhance tumor regions due to the contrast uptake in those locations and annul hypersignal regions present in the pre-contrast image. The process of subtraction was exclusively performed by Pelicano et al. [9]. For this project, the subtraction DCE-fl3D images, known as SUB-DCE-fl3D, were used, as they are particularly effective in revealing the entire tumor regions for tumor segmentation.

3.1   Pre-processing

The pre-processing applied to the breast MRI images followed the pipeline proposed by Pelicano et al. [9], which can be divided into three main steps: bias field correction, data normalization, and image filtering.

1. **Bias Field Correction.** The presence of bias field artifacts in MRI images can lead to unreliable intensity variations within voxels of the same tissue,

which can negatively impact the accuracy of intensity-based processing algorithms such as segmentation and classification. In order to address this issue and improve the reliability of the imaging data, the present study applied the SimpleITK N4BiasFieldCorrectionImageFilter algorithm to correct the bias field artifacts in all images [13].

2. **Data Normalization.** The images were normalized by scaling the voxel values between 0 and 255, using the Minimum-Maximum (Min-Max) normalization approach, which can be represented according to:

$$v' = \frac{v - min(A)}{max(A) - min(A)} (new_{max(A)} - new_{min(A)}) + new_{min(A)'} \qquad (1)$$

where v' and v represent the original and transformed values of each voxel; A is the volume data, with $max(A)$ and $min(A)$ being the highest and lowest value of A, respectively. The new maximum and minimum values for the transformed range are represented by $new_{max(A)}$ and $new_{min(A)}$, respectively [8].

3. **Image Filtering.** MRI images can be compromised by noise resulting from errors in image acquisition, which can negatively impact the performance of intensity-based segmentation algorithms. One type of noise commonly found in MRI images is Salt-and-Pepper noise, characterized by randomly scattered voxels that have been set to either 0 or the maximum intensity value [5]. In the present study, a median filter, which replaces the value of a voxel with the median gray level of its neighboring voxels, was applied to remove noise and also to smooth variations in signal intensity between tumorous and non-tumorous tissue, as suggested in [9].

3.2   Tumor Segmentation

As mentioned above, traditional methods to identify regions of interest in images, such as tumors, rely on overall intensity values but can be inaccurate when applied to heterogeneous tumors that have a range of intensity values, resulting in poor segmentation.

Thus, to address tumors with voxels spreading over a wide range of intensity values, a 3D region growing algorithm was used in this work. This algorithm operates by initiating with a seed point located within the structure of interest, which serves as the starting point for the growing process. Adjacent voxels with intensity values similar to that of the seed point within a defined threshold are subsequently added to the growing region through repeated evaluations of the intensity values of voxels in proximity to the region. The 3D region growing algorithm is a viable option for the segmentation of tumors in 3D images, as it is able to handle variations in intensity values within the structure, a characteristic commonly observed in tumors. However, it is important to note that the performance of the algorithm is dependent on how adequate the threshold value is. Additionally, the selection of a suitable seed point is a crucial aspect to ensure the proper functioning of the algorithm.

Threshold Selection. To determine the threshold for the application of the 3D region growing algorithm, a statistical analysis of the non-zero values in the 3D image was performed. Specifically, the mean and standard deviation were calculated. The threshold was then established as three standard deviations above the mean, as proposed by Pelicano et al. [9], as a mean to ensure that the threshold is set to a value that is sufficient to effectively differentiate the tumor from the surrounding tissue, while still being low enough to include all the voxels comprising the tumor within the same cluster.

Seed Selection. The initial stage of this tumor segmentation process involves the selection of seed points to start the segmentation. In [9], this process of seed selection was manual and only one single point was considered. However, in the case of multi-tumor images, multiple seed points may be necessary to capture all regions of interest. Furthermore, the heterogeneous nature of some tumors may result in inaccurate segmentation if relying solely on a single seed point based on the highest intensity. To address this limitation, an algorithm for automatic seed detection was developed.

1. Subsampling strategy

   In the first step, a subsampling strategy was employed. The image was divided into a grid of non-overlapping regions, by specifying a step size of 5 points in the x and y axis, resulting in a reduced set of points to be analyzed, rather than testing all points within the image. This 5-point step allowed a significant optimization of the computational efficiency, while still preserving the relevant information for the subsequent analysis.

2. Points of Interest

   In the second step, the algorithm examines each point in the reduced set, for each transverse cut, to determine if it is surrounded by a region of interest that can be segmented using the 3D region growing method. To do this, the algorithm evaluates whether the intensity of the surrounding voxels around the seed falls within the interval defined by the threshold. If this criterion is met, it may indicate the presence of a region of interest around the seed, and the point is considered for further processing in the algorithm. This step results in a set of points of interest that have been identified as having high-intensity values. However, it is important to note that these points may not necessarily correspond to regions of interest, such as tumors.

   Having identified the points of interest (x, y) of each transverse cut, the following processing steps were applied orderly to each one of the cuts. The order in which the cuts are processed was determined based on the number of points of interest identified in each cut, where cuts with higher number were processed first. This is because tumors are typically high-intensity structures, and a higher number of high-intensity points on a cut is indicative of a higher likelihood of finding a tumor. By prioritizing the processing of cuts with the highest number of points of interest, the algorithm is more efficient in identifying tumors.

3. Regions of Interest

   Having chosen a cut and its respective points, further analysis and evaluation are necessary to determine the true nature of these points of interest and distinguish them from other high-intensity zones of normal breast tissue. In order to differentiate between isolated high-intensity points and high-intensity regions, we employed a k-means clustering algorithm to group the high-intensity points according to their spatial location. The k-means algorithm resulted in multiple clusters, but not all the groups created by the algorithm necessarily correspond to regions of interest. To ensure that only regions of interest were being considered in the following analysis, we removed any clusters containing only a single point. This is because single points are more likely to be isolated high-intensity voxels, rather than part of a region of interest. As a result, only clusters containing multiple voxels were retained as potential regions of interest.

4. Region Segmentation

   To segment the regions of interest, the 3D region growing algorithm was applied to the centroid of each group. This resulted in a mask with all the regions of interest that had been segmented, in a three-dimensional representation. The segmentation process continued by analyzing the next transverse cut and identifying points of interest. To avoid duplicating efforts and optimize computational efficiency, if the points of interest were already included in the previous mask, they were not re-segmented. Instead, the algorithm focused on identifying new regions by applying 3D region growing to any points of interest that were not yet included in the mask. This process was repeated for all transverse cuts, ultimately resulting in a comprehensive mask that encompassed all structures identified across all cuts.

Structure Characterization. After conducting the segmentation and obtaining the resulting 3D structures, we performed an analysis of these structures by calculating their volume, circularity measures, and compactness.

1. Volume: The volume of the structures was determined by multiplying the voxel size ($1\,\text{mm} \times 1\,\text{mm} \times 1\,\text{mm}$) by the number of voxels in each structure. The resulting value represented the volume of the structure in mm^3. The volume determines the size of the structures, which can provide important information about their potential malignancy or benignancy.

2. Circularity: The circularity is based on the relationship between the volume and surface area of a structure, being calculated by multiplying the volume by 4π and then dividing it by the square of the surface area. The circularity value ranges between 0 and 1, with 1 indicating a perfectly circular structure and values closer to 0 indicating structures that are increasingly non-circular. This value can be used to characterize the shape of a structure, providing, in the context of medical imaging, information about its nature and behavior.

3. Compactness: The compactness of a structure is calculated as a measure of its shape, and is determined by the ratio of the volume of the structure to the volume of a sphere with the same surface area. As so, it can be calculated

by multiplying the square of the volume by 36π, and then dividing the result by the cube of the surface area. The compactness value ranges from 0 to 1, a value of 1 indicates a perfect sphere. Computing the compactness of each 3D structure can be useful in differentiating between different types of structures since tumors tend to have higher compactness and to be more encapsulated, while normal breast tissues tend to be more irregular in shape.

4   Results

The results of the implemented steps are demonstrated for a single MRI scan, containing one heterogeneous malignant tumor and four benign tumors, according to medical reports. The final segmentation of another MRI scan, which only contained one malignant tumor, was also presented.

ITK-SNAP (also known as the Insight Segmentation and Registration Toolkit), a powerful open-source software, which offers a wide range of tools for image analysis and visualization, was used in this study to visualize the MRI images of the breast, allowing for an enhanced view of the tumors present, in the transverse, coronal, and sagittal planes. Figure 1, on top, shows a transverse cut in which three different tumors can be seen - one malignant on the left, and two benign in the center and right. The images in the middle and bottom show the sagittal and coronal cuts, respectively, in which all tumors can be seen.

4.1   Image Filtering

The application of a median filter to the pre-processed image smooths the edges, facilitating the breast tumor segmentation process. Figure 2 illustrates the grayscale image before and after the filtering, on the left and right, respectively.

4.2   Tumor Segmentation

As mentioned above, for the automatic selection of the seed point, the proposed algorithm followed several steps. This subsection presents the results obtained in each step for one chosen transverse cut of the MRI exam containing multiple tumors.

After re-sampling the image in a reduced set of points, the algorithm will find, for each transverse cut, the points that are possible candidates as seeds. Figure 3, on the left, shows the location of the points of interest identified in the second step of the process for the transverse slice depicted in Fig. 2. The intensity of the voxels surrounding these points satisfies the established threshold criteria, thereby rendering them suitable as seeds. These points are characterized by high intensity, as they are represented in lighter shades of blue or yellow.

The subsequent step consisted of employing a selection criteria to differentiate between regions of interest and high-intensity isolated points. To achieve this, a k-means clustering algorithm was used. Figure 3, on the right, illustrates

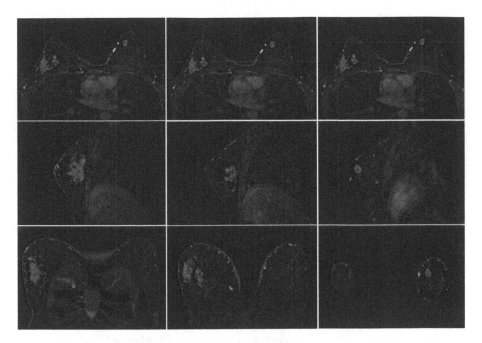

Fig. 1. Breast MRI visualization using *ITK-Snap*. The top images depict a transverse view, showcasing the presence of three distinct tumors. The middle and bottom images provide a clear representation of the same tumors in the sagittal and coronal planes, respectively. Left: malignant tumor; Center and Right: benign tumors.

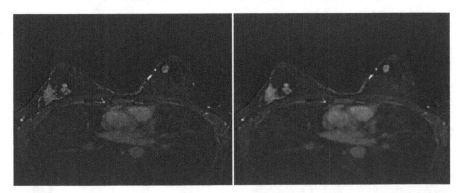

Fig. 2. Representation of the application of the median filter for edge smoothing: image prior to filtering (left) and image after filtering (right).

the clusters of high-intensity points obtained based on their spatial location, represented by different colors. Any cluster that consisted of a single point, such as the yellow cluster depicted in the figure, was subsequently disregarded.

In the following step, the 3D region-growing algorithm was applied using the centroid of the groups as seeds for the region-growing process. The intensity

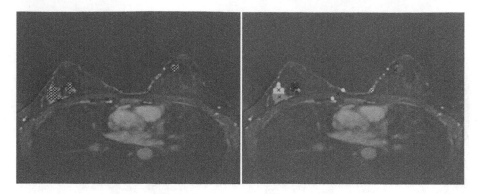

Fig. 3. Left: Location of the high-intensity points of interest for a transverse slice. Right: High-intensity points separated in groups according to their spatial location. The 'X' in black represents the centroids of the groups. (Color figure online)

threshold used was defined as mentioned above, i.e., three standard deviations above the mean intensity of the image. Figure 4, on the left, presents the resulting mask overlaid on the transverse cut. The mask is depicted in yellow, and a visual comparison with the structure of interest, highlighted in lighter blue in the original image, demonstrates a high degree of similarity. This suggests that the segmentation was successful. In Fig. 4, on the right, presents the mask obtained in a three-dimensional view, which further supports the accuracy of the segmentation, as it is observed that the structure of interest is well-segmented in three dimensions.

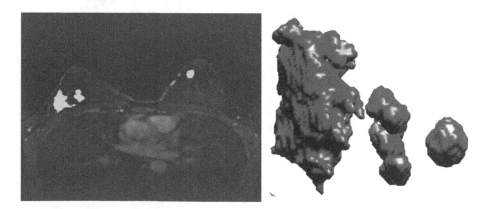

Fig. 4. Left: Output of the region growing algorithm applied to the centroids of the groups, in the previous transverse cut, represented in yellow. Right: Representation of the output of the region growing (yellow zones in a three-dimensional plane. (Color figure online)

Although these structures were identified only using seeds on one transverse cut, this process was successively applied to all cuts, which resulted in a total mask including regions of interest from all the cuts. As so, in Fig. 5 the resulting total mask displays four well-segmented structures that appear to be tumors, as well as one irregular structure that is not likely to be a tumor. This irregular structure is likely a segmentation artifact caused by high-intensity breast wall tissues. Furthermore, medical records indicated that this MRI exam included five tumors: one malignant and four benign. Thus, our methodology failed to segment one benign tumor - the smallest one, according to the records. This can be caused by limitations in the tuning of parameters, such as threshold or k-means parameters, in detecting small volumes. Additionally, Fig. 6 shows the mask obtained for the other MRI exam, in which our method did not perform well, despite the presence of only one tumor. The tumor's abnormally large size, irregularity, and heterogeneity made it challenging to segment, which suggests that further tuning of our parameters is required to improve performance in such cases.

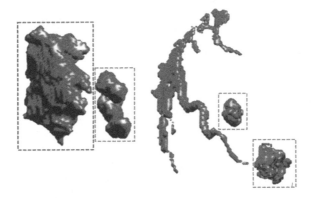

Fig. 5. Representation of the final 3D mask obtained when applying the 3D region growing successively in all the transverse planes of the multi-tumor MRI exam. Malignant tumors are identified with red, whereas benign tumors are identified with green. (Color figure online)

The masks obtained were divided into separate structures - five structures for the first exam, one single structure for the second exam - in order to be further analyzed. In Table 1, we present the values of volume, compactness, and circularity of each structure. For the first exam, structures are labeled from 1 to 5, according to their location in Fig. 5, from left to right.

Based on the results shown in Table 1, the malignant tumors were the structures with higher volumes, $11.682 \, mm^3$ and $12.259 \, mm^3$ for the first and second exams, respectively. They presented lower levels of circularity - 0.0067 and 0.0035 -, and also compactness - 0.1502 and 0.0586 -, when compared to the benign tumors, which indicates that their structures are more irregular and less round.

Fig. 6. Representation of a poorly-segmented tumor obtained when applying the proposed methodology to one single-tumor MRI exam.

Table 1. Characterization of the structures according to their volume, compactness, and circularity. The classification was based on medical records and human inspection.

Structure	Volume	Compactness	Circularity	Classification
1–1	11682	0.1502	0.0067	Malignant
1–2	1938	0.2016	0.0148	Benign
1–3	845	0.0167	0.0037	Not Tumor
1–4	763	0.8451	0.0526	Benign
1–5	1813	0.2602	0.0180	Benign
2–1	12259	0.0586	0.0035	Malignant

The three benign tumors from the first exam (structures 2, 4, and 5) presented the three highest values of compactness and circularity, highlighting their round compact shape. The third structure, which is not a tumor, has a significantly lower compactness value compared to the other structures - 0.0167 - and also exhibits the second lowest values for both volume and circularity, 845 mm^3 and 0.0037, respectively.

The results align with the typical characteristics of malignant and benign tumors, providing additional information that could be used to improve the effectiveness of our pipeline at identifying and discarding non-tumor structures.

5   Discussion

Previous algorithms, such as the one proposed by Pelicano et al. [9], used the highest intensity voxel as the seed for the region growing algorithm. However, this approach may not always correctly identify the tumor region and may require manual selection of the seed. Additionally, it only allows selecting one seed, which prevents targeting more than one region. Our proposed algorithm overcomes these limitations by utilizing a k-means clustering approach to automatically select multiple seeds for the region growing algorithm, providing a more robust, efficient and automatic method for multi-tumor segmentation.

The results of the proposed methodology for tumor segmentation suggests a promising performance, as indicated by visual inspection. At least, four tumors appear to have been segmented with high accuracy. However, due to the absence of actual tumor structures for comparison, a comprehensive evaluation of the effectiveness of the method is challenging. Nonetheless, the results partially agree with medical records. In the first examination, the algorithm successfully segmented a structure that displayed characteristics commonly associated with malignant tumors, such as an irregular shape and a larger size. The structure identified in the second exam, which is also a malignant tumor, also presented a larger size and irregular shape. Additionally, the algorithm identified three structures in the first exam with round regular shapes and smaller sizes, which agree with the characteristics of benign tumors. Based on this visual analysis, it can be inferred that the proposed pipeline demonstrated to be promising in segmenting both malignant and benign tumors.

However, there were some drawbacks during the segmentation process. The segmentation of the first MRI scan revealed two more shortcomings: one benign tumor was missed and one irregular structure that was not a tumor was mistakenly identified. The limitations of the segmentation process may stem from the parameters chosen throughout the pipeline. The threshold used was established in [9], but distinct results could have been achieved with different thresholds. The threshold is critical during the 3D region growing process, as it determines the range of intensities around the seed point that is considered for segmentation. Thus, if too low, it may lead to the inclusion of regions with a low intensity that are not tumors, while a high threshold may result in the exclusion of high-intensity regions, that can be in fact tumors.

The k-means clustering results may have been influenced if different parameters for k had been used. In this case, the k-means was set to create 6 groups, and groups with only one point were not further considered for segmentation. However, other k values would have changed the results obtained. For instance, if the k parameter was changed from 6 to 8, the seed points identified would be different, and this would lead to a distinct resulting mask. The condition that single-point groups should not be considered helps remove high-intensity points that are not part of a tumor and may lead to removing points of interest that actually belong to smaller regions of interest. This may have caused the missing segmentation of the small benign tumor in the first image.

The parameters used in the algorithm play a crucial role in determining the results obtained, as changes made to any parameter will have an impact on the outcome. Hence, to improve the effectiveness of the proposed methodology, we recommend performing a more thorough analysis of the threshold and k-means clustering tuning parameters in future work.

In order to further analyze the structures identified by the algorithm, various features were computed, including volume, compactness, and circularity. These features offer valuable insights into the structural attributes, despite the absence of precise threshold values specific to each tumor type. Nevertheless, it is anticipated that malignant tumors will exhibit lower levels of compactness and circularity, attributed to their irregular shapes, along with larger volumes, as they typically manifest in larger sizes. The analysis of the results showed that, in fact, the structures corresponding to the malignant tumors presented higher volumes and lower compactness and circularity compared to the benign tumors, as expected. The structures identified as benign tumors had a more round and compact shape, as well as smaller volume, which is also consistent with the expected characteristics of benign tumors. The non-tumor structure had lower compactness compared to the tumor structures and was also characterized by a low volume and circularity.

These findings provide additional information that could be used to differentiate between malignant and benign tumors, as well as tumors and normal breast tissue. In the future, by integrating these features with features of texture and intensity, into a machine learning algorithm, the segmented structures could be classified as malignant, benign, or non-tumor.

6   Conclusion

In conclusion, our study aimed to develop a fully automated methodology for multi-tumor segmentation in breast MRI. Our proposed pipeline successfully achieved both the completely automatic seed selection and the multi-tumor segmentation objectives, leading to a successful segmentation in the majority of the cases. However, there are some limitations to the pipeline, such as difficulties in segmenting certain tumors, which suggest that future work should focus on finding optimal parameters for the model. Additionally, by integrating features such as volume, compactness, and circularity into machine-learning algorithms, the segmented structures could potentially be classified as malignant, benign, or non-tumor. Moreover, our method was evaluated on a limited sample size, consisting of only two MRI exams from two different patients. This hinders the generalization of the results to a larger population. As such, future research should aim to expand the sample size and test the proposed pipeline on a more representative population in order to validate the results.

Overall, our study presents a promising approach for automatic multi-tumor segmentation in breast MRI and holds potential for refinement through the optimization of model parameters and integration with a classification model based on the characteristics of the segmented structures.

Acknowledgments. This work was funded by national funds through FCT - Fundação para a Ciência e a Tecnologia, I.P., under the PhD grant UI/BD/153605/2022 (T.M.C.P.), the Scientific Employment Stimulus CEECIND/03986/2018 (R.S.) and CEECINST/00013/2021 (R.S.), and within the R&D units IEETA/UA UIDB/00127/2020, IBEB UIDB/00645/2020, 2022.08973.PTDC, and CICECO-Aveiro Institute of Materials UIDB/50011/2020, UIDP/50011/2020, and LA/P/0006/2020 (PIDDAC).

References

1. Global Cancer Observatory. International Agency for Research on Cancer - World Health Organization. https://gco.iarc.fr/. Accessed 12 Dec 2022
2. Al-Faris, A.Q., Ngah, U.K., Isa, N.A.M., Shuaib, I.L.: Computer-aided segmentation system for breast MRI tumour using modified automatic seeded region growing (BMRI-MASRG). J. Digit. Imaging **27**(1), 133–144 (2014). https://doi.org/10.1007/s10278-013-9640-5
3. Baccouche, A., Garcia-Zapirain, B., Catillo Olea, C.: Connected-unets: a deep learning architecture for breast mass segmentation. NPJ Breast Can. **7**(151) (2021). https://doi.org/10.1038/s41523-021-00358-x
4. Chen, D.R., Chang, Y.W., Wu, H.K., Shia, W.C., Huang, Y.L.: Multiview contouring for breast tumor on magnetic resonance imaging. J. Digit. Imaging **32**(91) (2019). https://doi.org/10.1007/s10278-019-00190-7
5. Gonzalez, R., Woods, R.: Digital Image Processing. Prentice Hall, Hoboken (2002)
6. Huang, C.L.: Breast mass segmentation on breast MRI using the shape-based level set method. Biomed. Eng. Appl. Basis Commun. **26**(4), 1440006 (2014). https://doi.org/10.4015/S1016237214400067
7. Melouah, A., Layachi, S.: A novel automatic seed placement approach for region growing segmentation in mammograms (2015). https://doi.org/10.1145/2816839.2816892
8. Patro S., Sahu K.: Normalization: a preprocessing stage. arXiv (2015). https://doi.org/10.48550/arXiv.1503.06462
9. Pelicano, A.C., et al.: Development of 3D MRI-based anatomically realistic models of breast tissues and tumors for microwave imaging diagnosis. Sensors **21**(24), 8265 (2021). https://doi.org/10.3390/s21248265
10. Rangayyan, R., El-Faramawy, N., Desautels, J., Alim, O.: Measures of acutance and shape for classification of breast tumors. IEEE Trans. Med. Imaging **16**(6), 799–810 (1997). https://doi.org/10.1109/42.650876
11. Shrivastava, N., Bharti, J.: Breast tumor detection and classification based on density. Multimed. Tools Appl. **79**, 26467–26487 (2020). https://doi.org/10.1007/s11042-020-09220-x
12. Thakran, S., Chatterjee, S., Singhal, M., Gupta, R., Singh, A.: Automatic outer and inner breast tissue segmentation using multi-parametric MRI images of breast tumor patients. PLoS ONE **13**(1), e0190348 (2018). https://doi.org/10.1371/journal.pone.0190348
13. Tustison, N.; et al.: N4ITK: improved n3 bias correction. IEEE Trans. Med. Imaging **29**(6), 1310–1320 (2010). https://doi.org/10.1109/TMI.2010.2046908
14. Wang, Z., Qu, Q., Yu, G., Kang, Y.: Breast tumor detection in double views mammography based on extreme learning machine. Neural Comput. Appl. **27**, 227–240 (2016). https://doi.org/10.1007/s00521-014-1764-0
15. Wang, J.; Yang, Y.: A context-sensitive deep learning approach for microcalcification detection in mammograms. Pattern Recogn. **78**(6) (2018). https://doi.org/10.1016/j.patcog.2018.01.009
16. Yin, D., Lu, R.W.: A method of breast tumour MRI segmentation and 3D reconstruction. In: 2015 7th International Conference on Information Technology in Medicine and Education (ITME), pp. 23–26 (2015). https://doi.org/10.1109/ITME.2015.117

Mortality Prediction via Logistic Regression in Intensive Care Unit Patients with Pneumonia

Nuno Pedrosa[1] and Sónia Gouveia[1,2]([✉])

[1] IEETA-Institute of Electronics and Informatics Engineering of Aveiro,
DETI-Department of Electronics, Telecommunications and Informatics,
University of Aveiro, Aveiro, Portugal
[2] LASI-Intelligent Systems Associate Laboratory, Portugal, University of Aveiro,
Aveiro, Portugal
sonia.gouveia@ua.pt

Abstract. This work focuses on the problem of mortality prediction in patients with pneumonia after admission into an intensive care unit, by addressing it via logistic regression. This approach can model the relationship between clinical correlates and the probability of the binary outcome, with obvious advantages such as simplicity and interpretability of the predictive models. This work further inspects the potential of localized models, an approach based on different (parallel) predictive models each one constructed in clusters automatically identified in the training set. The predicted outcome is then obtained via membership separation (M, which corresponds to the outcome of the closest localized model) or weights (W, outcome as the weight average of localized outcomes via inverse distance). The results point out a similar balanced accuracy of 0.73 for the global model M24-48PS (without oversampling) and the W M24-48PSC model (weighted average of localized models without oversampling), which is partially explained by the small separability between the identified clusters. Therefore, a proof of concept was performed to support the usefulness of localized models in more separable data. This study considered a small amount of data for training and testing (chosen as that closest to the centroids of the identified clusters) and the results suggest that the localized approach can outperform the global one in more separable data.

Keywords: Pneumonia Mortality · Machine Learning · Logistic Regression · Localized Models

1 Introduction

Pneumonia constitutes a serious illness that can lead to critical medical complications, or even death, being characterized by inflammation of the lungs, mainly affecting the lung alveoli [21]. Pneumonia is one of the most common infectious causes of death worldwide and is the fourth major cause of death after heart disease, stroke, and chronic obstructive pulmonary disease. The Global Burden of Disease estimates that pneumonia caused 2.5 million deaths worldwide in

© Springer Nature Switzerland AG 2024
V. Vasconcelos et al. (Eds.): CIARP 2023, LNCS 14470, pp. 30–44, 2024.
https://doi.org/10.1007/978-3-031-49249-5_3

2019 across all age groups [10]. In outpatient contexts, especially in developed countries, pneumonia has a relatively low fatality rate ranging from 1% to 5%. However, among patients who require hospitalization, the mortality rate is much higher and reaches up to 25%, particularly when admission to an ICU (intensive Care Unit) is necessary [12].

The cases of Community Acquired Pneumonia (CAP, the most common type of pneumonia) range from 1.5 to 14 cases per 1000 persons per year, with the value of a specific region depending on geography, wealth and population characteristics [24]. At any level of medical care, treating patients with CAP accompanies high direct and indirect costs. As an example, the costs in USA range from 8.4 to 10 thousand million dollars annually [19], with a cost per patient that required hospitalization of 10 thousand dollars (from 2008 to 2014) [28]. In Europe, the costs are about 10.1 thousand million euros annually [32]. All of this with an increasing trend in costs aggravated by inflation [1]. This makes pneumonia not only a health problem, but also a financial burden.

Predicting outcomes related to pneumonia is crucial from a medical and financial perspective. In this work, the focus is on a distinct prediction that can be simplified into a binary problem, the risk of mortality following admission to ICU (Intensive Care Unit). It is a big challenge to predict because ICU mortality rates range greatly depending on various patient characteristics, including physiologic characteristics, clinical, and demographic [15]. There are several advantages to predicting mortality in a hospital setting. Firstly, it can help determine the severity of an illness, as well as the potential effectiveness of advanced therapies [23]. Secondly, it can aid in triaging and allocating resources, which can be critical in times of high demand or resource scarcity. Additionally, mortality prediction can be useful for health research and administration, for benchmarking and evaluating the performance of healthcare systems in comparison to observed death rates. Lastly, it can also facilitate conversations with patients and their families about predicted outcomes and inform healthcare policies [15].

There are already some clinical scores that are commonly used as daily tools in the hospital setting to predict mortality after admission to the ICU, including APACHE II (Acute Physiology and Chronic Health Evaluation II) [13], SAPS II (Simplified Acute Physiology Score II) [11] and SOFA (Sequential Organ Failure Assessment) [29]. For mortality specific to patients with pneumonia, there is CURB-65 [18] and PSI/PORT (Pneumonia Severity Index) [7]. Although useful, alternative methods to predict mortality can substantially improve accuracy and simplicity. Evaluating a high number of scores can make the process more complicated, and some scores may not be adapted to specific situations. By using alternative methods, predictions can be more specific and accurate.

ML (Machine Learning) methods have gained considerable attention in the medical field due to their excellent performance in various prediction tasks. These methods have the potential to enhance diagnosis and outcome prediction, particularly as the number of healthcare providers using electronic records continues to increase [27]. The predictive objectives of this study align with this perspective, and thus, ML methods will be employed to predict the possibility of mortality after admission to the ICU. This topic has already been studied in the literature using various approaches, with some of the more recent summarized in Table 1.

Table 1. Recent research papers. Acronymous: GLM - Generalized Linear Model, CPN - Causal Probabilistic Network, XGBoost - Extreme Gradient Boosting, GBM - Gradient Boosting Machine, MLP - Multilayer Perception, KNN - K Nearest Neighbours, NN - Neural Networks, SVM - Support Vector Machine, LR - Logistic Regression.

Method	Dataset and Details	AUC	Ref.
SeF-ML (CPN to predict mortality in sepsis, adapted with ML).	From two Spanish university hospitals. With 4,531 CAP patients in total. With Structured health data. For 30-day CAP mortality.	0.801	[4]
Extensive-CURB-RF model (CURB-65 and Random Forest mix model).	From an emergency department in Seoul, Korea with 1,732 patients. With demographic information, mental status and laboratory findings, including CURB-65 variables. For 30-day pneumonia mortality.	0.822	[12]
Random forest (best), LR (also best), GBM and XGBoost with SMOTE.	From humanitas Research Hospital in Milan, Italy. With 1,135 consecutive patients with COVID-19. With vital parameters, laboratory values and demographic features.	0.88 (best)	[14]
XGBoost (best), Data Mining, Gaussian Naive Bayes, KNN and SVM.	From a Hospital in Cuenca, Spain. With symptom and vital signs, analytical parameters, chest X-ray, demographic characteristics and presence of comorbidities. For 150 COVID-19 patients.	0.93 (best)	[2]
XGBoost for different time intervals.	MIMIC dataset, with 60,000 ICU admissions with demographics, vital signs, laboratory tests, medications and caregiver notes.	0.87–12h 0.78–24h 0.77–48h 0.73–72h	[26]
XGBoost (best), GLM, Random Forest and NN (also best)	From german Helios hospitals with 241,988 patients, with administrative hospital data. For Severe acute respiratory infections.	0.834 (best) 0.830 (GLM)	[16]
LR , Random Forest, SVM, LightGBM (best), MLP and XGBoost	52,626 patients with pneumonia in Taiwan between 2010 and 2019, with 33 features including vital signs, laboratory data and underlying comorbidities.	0.835 (best) 0.807 (LR)	[3]

Table 1 shows that recent studies have achieved a fair performance while using a diverse range of data, with different origins, sizes and types of patients. It also presents a tendency for the use of complex models, with special emphasis on XGBoost. Although the use of complex models can result in good predictions, the use of simpler and interpretable models, such as logistic regression, enables to better understand the rules and correlations of variables in a model. Another important aspect that is often overlooked is the significant variation that exists within the populations from which the data originates. Different subsets of the population possess unique characteristics that can have varying impacts on mortality predictions. To address this, one approach is to partition the data into more homogeneous populations and develop separate models for each population group.

This work has the objective to predict, with machine learning, the probability of mortality in patients with pneumonia after entrance in ICU. This will be done by resorting to hospital data obtained from patients with pneumonia in ICU. Another objective is to separate the data into different more homogeneous populations, and then obtain different models for the different populations, in order to try to increase the performance of the predictions. All of this with the aim to improve outcome prediction in a hospital environment, with great importance in terms of hospital logistics, ensuring health care quality.

2   Materials and Methods

This work is based on experimental data collected from patients admitted into a ICU, in an anonymous and confidential manner. The methods used in this work aimed at data pre-processing and modelling, which are described in more detail in the following subsections.

The data preprocessing and modelling were carried out using Python, highlighting the use of the following packages. SciPy in the Yeo and Johnson algorithm for the transformation of the data and the linkage algorithm to construct the dendrograms to separate clusters for localized models and to select highly correlated features [30]. Scikit-learn for the Standard scaling, for the imputation related functions (KNN Imputer, Iterative Imputer, IterativeImputer, Random Forest Regressor and Bayesian Ridge) and for the logistic regression implementation [22]. Imbalanced-learn in the oversampling technique ADASYN [17]. Finally, Pandas and NumPy for general data manipulation [8, 20].

2.1   Experimental Data

The data under analysis contains information on a large number of patient admissions in the ICU diagnosed with pneumonia, in Portugal, from February 02, 2009 to August 18, 2020 in about 5000 health units. The data were simulated from the real dataset to ensure the complete anonymity of the patients and non disclosure of the real data, using the joint probability distributions empirically estimated from the real data. The data simulated is confidential under a signed agreement between the involved parties in order to make sure that the information conveyed in the data is not released to the general public. This complies with the overall rules of the General Data Protection Regulation (GDPR) requirements on collecting, storing and managing personal data.

The data comprised a total of 15355 admissions mimicking a total of 7719 unique patients. The variables available included the associated with the hospital episode, clinical and logistic data e.g. patient age, birth date, gender, vital signs, laboratory measurements, diagnoses, medications, clinical devices, types of services and others. In order to extract the variables from the database two criteria were used, firstly, variables similar to those found in the literature were sought, and then, other factors were extracted based on human intuition and later analyzed their importance for solving the problem. The chosen time interval of a patient's stay in the hospital to extract starts at 24h and ends at 48h after the patient's entrance into ICU, to do predictions based on the 24–48h patient status. After pre-processing, the sample consisted of a set of 64 features observed from 2729 patients, with a final mortality ratio of 17.3%.

2.2   Data Pre-processing

The pre-processing tasks were divided into 7 steps, including, (1) transformation, (2) outlier removal, (3) first feature/observation removal, (4) scaling, (5) imputation, (6) selection and (7) second feature/observation removal.

In the first step, the Yeo and Johnson transformation was used to reduce asymmetry in the distribution of the data [33]. Equation 1 presents the transformation, where y is the original value and λ is a chosen parameter optimized as to provide the best approximation to a normal distribution.

$$y(\lambda) = \begin{cases} \frac{(y+1)^{\lambda}-1}{\lambda} & \text{if } \lambda \neq 0,\, y \geq 0 \\ log(y+1) & \text{if } \lambda = 0,\, y \geq 0 \\ \frac{-((-y+1)^{2-\lambda}-1)}{2-\lambda} & \text{if } \lambda \neq 2,\, y < 0 \\ -log(-y+1) & \text{if } \lambda = 2,\, y < 0 \end{cases} \tag{1}$$

The choice of the optimal λ is performed via log-likelihood maximization. The transformation is applied to each feature (separately) with a skewness greater than 1 in absolute terms. Here, features with a negative skew were previously changed into a positive skewed feature, where the new value of each observation corresponds to the max value of the variable minus the value of the observation.

The second step consisted in outlier removal based on a rule constructed from the empirical IQR (Interquartile Range) multiplied by a fairly large constant in an effort to just eliminate very severe outliers. This work considered a multiplication factor of 5.

Third, in first feature/observation removal, features and observations with a big amount missing data (NaNs) were eliminated. The thresholds of elimination were more than 8000 NaNs for features and 23 for observations. These values were chosen in an attempt to balance the loss of information with the elimination of observations and features.

Fourth, in scaling, the data was scaled using Standard scaling. It adjusts each feature to have a variance of one while removing the mean. This is important to make features comparable to the model.

Fifth, in imputation, data imputation is used to estimate the values of NaN data. Three different imputation techniques were used based on two imputation functions, KNN Imputer and Iterative Imputer [22]. Regarding IterativeImputer, 2 techniques were used with different estimators, Random Forest Regressor and Bayesian Ridge. To evaluate and compare the imputation results, the R^2 (coefficient of determination) mean values were calculated based on validation data, where 200 available data points were randomly selected and replaced with NaN values in 5 validation iterations. This was done for each function and feature. In the end, for each feature, the technique with the best R^2 is saved, and if the R^2 is longer than 0.3, the imputed feature will substitute the original feature.

Sixth, in selection, the influence of each feature in logistic regression becomes less precise if the input variables have a high degree of correlation between them. This is known as multicollinearity. To remove highly correlated features, a correlation dendrogram analysis was performed based on dissimilarity, which corresponds to 1 minus the absolute value of the correlation. The threshold of dissimilarity to separate highly correlated features from non-correlated is 0.3 [6]. Then from a pull of correlated features, the one to keep is the one with a higher correlation with the prediction objective variable.

Seventh and final, in the second removal, the remainder of NaN data was eliminated. First, features with a larger fraction of NaNs than 0.3 were eliminated. Then, the rows that still had NaNs were eliminated, achieving the final 2729 admissions and 64 features.

2.3   Predictive Models

This work makes use of Logistic Regression, which is characterized by a smooth and linear decision boundary between two classes, making it good for linearly separable data and permitting binary classification [5]. It constitutes a simple model that is easy to interpret and fast to train. Also, Recursive Feature Elimination (RFE) based on logistic regression was used to choose the most important features to use in the final models.

Oversampling techniques permit to improve the performance of models in unbalanced data. ADASYN (Adaptive Synthetic Sampling Method for Imbalanced Data) is the oversampling technique used in this work. It allows to deal with unbalanced datasets by creating synthetic data to balance the classes, and thus obtain better results in the algorithms. ADASYN works similarly to the popular algorithm SMOTE, but skews slightly the sample space to the points that are not located in the homogenous neighbourhoods, adding a small random value, making the data more realistic [9].

Twelve different models were created, namely, 2 global models, denoted by M24-48PS (without oversample) and OSM24-48PS (with oversample) and 10 localized models, the M24-48PSC family (without oversample) and the OSM24-48PSC family (with oversample). The names of the models were created according to their specifications, the core, M24-48PS, stands for mortality given the 24–48h patient status, the OS, in the beginning, for oversampled and the C, in the end, for clustered. A simplified pipeline of the procedure to construct the models is presented in Fig. 1.

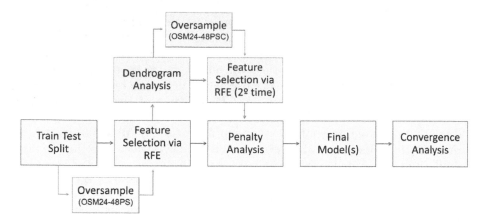

Fig. 1. Pipeline to obtain the models. Blue squares are in common for all models, green squares are just for the identified OS model and orange squares are just for C models. (Color figure online)

A more detailed explanation of the steps present in Fig. 1 are as follows.
Train test split: To split the data, a 0.33 ratio was used, resulting in 1828 observations (1512 in class 0 and 316 in class 1) for train and 901 observations (745 in class 0 and 156 in class 1) for test. Class 1 stands for the mortality class and class 0 for the non-mortality one. Oversample (OSM24-48PS): The data for the global oversampled model (OSM24-48PS) is oversampled, balancing the – classes.

- Feature selection via RFE (Recursive Feature Elimination): RFE was applied and 20 features were obtained that make a substantial difference in the models performance, consequently, these are the features that will be used.
- Dendrogram analysis (for localized models): To separate the data into different populations, hierarchical clustering dendrogram analysis was applied to the train data. The clustering was done with the Ward variance minimization algorithm and the Euclidean metric [31]. It got 5 clusters as the optimal separation, with a silhouette score of 0.11 [25]. The low silhouette value shows that some cluster overlap is to be expected, however, it was still possible to obtain different clusters, all of them with a different number of observations and mortality ratios.
- Oversample (OSM24-48PSC): Each cluster is oversampled independently for the OSM24-48PSC family of models data to balance the classes.
- Feature selection via RFE (second time for localized models): This time, it was done for each specific cluster, because important features in a single cluster can differ from the important features in all the data.

- Penalty analysis: The performance of the models with different penalties was accessed (L1, L2 and Elastic Net with various ratios of L1 and L2 penalties, according to the implementation of logistic regression in Scikit-learn [22]). All the models got L2 (standard one) as the final one, because the results with other penalties were not significantly different.
- Final model(s) and convergence analysis: The final models were obtained and the convergence of the cost functions and parameters were confirmed.

3   Results and Discussion

The final models are summarized in Table 2, there, chosen features for each model have the coefficient value, with the total number of features chosen for a model present in the Total row. The number of observations for training and the corresponding mortality ratios are presented in the following rows. Some features have their name abbreviated. Their meaning is the following, TAD is diastolic blood pressure, PaCO2 is the partial pressure of oxygen, NIVUD is non invasive ventilation until date, MultiorganD is Multiorgan Dysfunction, NeuroMyopathy is Polyneuropathy/Myopathy, metHB is Methemoglobin, AST/TGO is Aspartate Aminotransferase, DHL is Lactate Dehydrogenase, GGT is Gamma-Glutamyl Transferase and BE is the Base Excess. Some relevant aspects that can be highlighted from Table 2 are the great variety of different mortality ratios in the different clusters, evidencing that different populations in the data have different probabilities of mortality. The great variation in the number of observations for each population. And the fact that each cluster had an optimal pull of different features shows that important factors to predict mortality vary from population to population.

Examining the coefficient values present in Table 2, that permit to assess the importance of features in each model, M24-48PS and OSM24-48PS models exhibit similar coefficient patterns. The ones that lead larger increase in the probability of mortality (larger positive value) are DHL, PaCO2 and GGT and the ones that lead to a smaller probability of mortality are TAD, BE, AST/TGO, AlkalinePhosph and Age (the true age leads to a larger probability of mortality, because it was inverted in pre-processing). In the coefficients of the localized models, there is considerable variation in the importance of the features, indicating that different populations exhibit distinct mortality indicators, and that the importance of these indicators varies across clusters. These insights are very useful in a hospital environment and help doctors understand how a prediction of a model was made, and change their procedures accordingly.

Table 2. Models Summary, with the features used in each model with their coefficient value, number of train observations used and the mortality ratios. Some features have their name abbreviated.

Parameters	M24-48PS	OSM24-48PS	M24-48PSC					OSM24-48PSC				
			1	2	3	4	5	1	2	3	4	5
Age	-0.38	-0.40	0.56		-0.21	-0.71	-0.46	0.88	0.01	-0.31	-0.81	-0.58
HeartRate	0.19	0.23	0.53	0.67	0.31		0.13	1.09	0.75	0.25		
TAD	-0.29	-0.37		-1.25	-0.36		-0.26	-1.48	-1.67	-0.31		-0.26
Glasgow	0.26	0.27			0.22		0.35		-0.29	0.23		0.36
PaCO2	0.44	0.51			0.42	0.70	0.38	0.58	0.84	0.50	0.79	0.51
NIVUD	-0.23	-0.28			-0.16		-0.13	-0.96	-0.31	-0.23	0.77	-0.30
Pneumothorax	0.16	0.15			-0.01				-0.01			
MultiorganD	0.29	0.20		0.61	0.22				0.55	0.12		
NeuroMyopathy	-0.19	-0.16			-0.18		-0.05		0.43	-0.07		
Shock	0.25	0.20			0.07		0.02	-0.19	-0.45			
Lactate	0.24	0.31		0.66	0.29		0.21	-0.42	0.60	0.37		0.30
metHb	0.20	0.24		-0.62	0.15	0.48	0.29	-1.13	-0.74	0.17	0.73	0.35
AST/TGO	-0.34	-0.44			-0.27		-0.34	-0.80	0.12	-0.39		-0.48
DHL	0.63	0.66		-0.51	0.76	0.65	0.60	0.05	-0.57	0.89	0.88	0.65
AlkalinePhosph	-0.22	-0.33			0.16		-0.45	0.22	0.59	0.06		-0.52
GGT	0.36	0.47			-0.03	0.38	0.42	0.81	0.23			0.61
Platelets	-0.24	-0.30			-0.07		-0.31	-0.58	-0.90	-0.13		-0.40
Urea	0.22	0.19		0.99	0.11		0.24	0.30	1.26			
Lymphocytes	-0.15	-0.22			0.09		-0.28	-0.49	0.24	0.09		-0.33
BE	-0.36	-0.37		-0.60	-0.33	-0.68	-0.31	-0.46	-1.02	-0.28	-0.80	-0.38
Total	20	20	2	8	20	6	18	16	20	16	6	14
Observations	1828	2934	69	45	294	358	1062	126	59	360	634	1848
Mortality Ratio	17%	48%	8%	36%	38%	10%	14%	50%	49%	50%	49%	50%

3.1 Model Performance and Comparison

The performance via cross-validation with 10 iterations for each model is present in Table 3. There, the localized models and global models have approximately the same accuracy and balanced accuracy, suggesting that the data may not exhibit sufficient heterogeneity to yield significantly improved predictions with the localized models, evidenced by the low silhouette score. Furthermore, it is noteworthy that the oversampled models consistently outperform the non-oversampled models across various performance metrics (except for accuracy, expected due to data imbalance). In addition, the Recall, and subsequently F1-score, are especially low for non oversampled models, implying that the percentage of data samples that the model correctly identified as 1 in mortality is low. In terms of specific localized models, significant variations in performance across different cluster models are observed, pointing to the obtainment of different models with distinct predicting capabilities. Lastly, it is important to note that the M24-48PSC1 model exhibits Precision, Recall, and F1-score values of 0, because the performance functions had a zero division due to the low amount of positive mortality predictions present in the train data. In such cases, the function returns 0.

Table 3. Performance metrics via cross-validation for all models. W. Average is the weighted average of the specific localized models, weighted according to the amount of train data for each model.

Model	Accuracy	Balanced Accuracy	Precision	Recall	F1-score
Global Models					
M24-48PS	0.85 ± 0.02	0.62 ± 0.04	0.65 ± 0.16	0.28 ± 0.09	0.38 ± 0.11
OSM24-48PS	0.72 ± 0.03	0.72 ± 0.03	0.72 ± 0.03	0.70 ± 0.06	0.71 ± 0.04
M24-48PSC Localized Family					
W. Average	0.85 ± 0.03	0.61 ± 0.07	0.60 ± 0.26	0.25 ± 0.11	0.33 ± 0.14
1	0.91 ± 0.07	0.70 ± 0.24	0.00 ± 0.00	0.00 ± 0.00	0.00 ± 0.00
2	0.82 ± 0.16	0.81 ± 0.19	0.72 ± 0.33	0.75 ± 0.34	0.71 ± 0.31
3	0.68 ± 0.07	0.65 ± 0.08	0.62 ± 0.16	0.49 ± 0.09	0.55 ± 0.11
4	0.90 ± 0.04	0.57 ± 0.11	0.40 ± 0.49	0.15 ± 0.21	0.21 ± 0.2
5	0.88 ± 0.02	0.59 ± 0.04	0.69 ± 0.22	0.21 ± 0.07	0.31 ± 0.10
OSM24-48PSC Localized Family					
W. Average	0.73 ± 0.05	0.73 ± 0.05	0.73 ± 0.05	0.74 ± 0.10	0.73 ± 0.06
1	0.81 ± 0.09	0.81 ± 0.09	0.78 ± 0.13	0.92 ± 0.13	0.83 ± 0.08
2	0.81 ± 0.16	0.81 ± 0.16	0.78 ± 0.17	0.90 ± 0.21	0.82 ± 0.18
3	0.65 ± 0.06	0.65 ± 0.06	0.66 ± 0.09	0.62 ± 0.08	0.64 ± 0.07
4	0.77 ± 0.09	0.76 ± 0.09	0.76 ± 0.08	0.76 ± 0.12	0.76 ± 0.09
5	0.73 ± 0.03	0.73 ± 0.03	0.73 ± 0.03	0.74 ± 0.09	0.73 ± 0.04

The performance with test data is present in Table 4. There, to be able to compare the test results of the global and localized models, two systems of assigning test data to the localized models were used. First, membership Separation (M), where the observations of the test data are assigned to the cluster with a smaller distance to the cluster centroid. Second, via weights (W) were all the test data is predicted using all models, and the final probability for a given observation is the weighted average over all model predictions. The weights are calculated via inverse distance weighting (IDW) between the observation and the cluster centroids.

In Table 4, global and localized models do not have a very significant difference between them in performance, consistent with the findings from cross-validation. However, contrary to the cross-validation results, the models trained with oversampled data show slightly lower performance compared to the models without oversampling. This suggests that the performance improvement observed with oversampling in cross-validation did not generalize well to the test data with the original mortality ratios. By comparing specific localized models, there is a trend were models trained for the same cluster, with or without oversample, have the same tendency to be better or worse compared to the other models. This indicates that certain clusters have inherent factors that make it easier or more challenging for the models to accurately predict outcomes. Notably, the models obtained with cluster 1 exhibit the poorest performance, likely due to the small number of observations in its training data and the low probability of mortality in that cluster. Conversely, the models built with cluster 2 demonstrate the best performances. Finally, out of all the models, M24-24PS and M24-24PSC Weights can be picked as slightly better in terms of performance in test data.

Table 4. Performance metrics obtained with test data. Individual localized models are evaluated via membership. Membership A. is the weighted average of the localized models membership, weighted according to the amount of test data for each model.

Model	Accuracy	Balanced Acc.	Precision	Recall	F1-score	AUROC	Op. Threshold
Global Models							
M24-48PS	0.73	0.73	0.36	0.74	0.49	0.78	0.15
OSM24-48PS	0.69	0.72	0.33	0.76	0.46	0.77	0.43
M24-48PSC Localized Family							
Weights	0.72	0.73	0.35	0.74	0.48	0.78	0.18
Membership A.	0.67	0.72	0.32	0.78	0.44	-	-
1	0.58	0.76	0.22	1.00	0.36	0.63	0.07
2	0.81	0.86	0.62	1.00	0.77	0.85	0.30
3	0.72	0.69	0.54	0.63	0.58	0.73	0.41
4	0.58	0.69	0.25	0.86	0.38	0.75	0.07
5	0.69	0.72	0.27	0.77	0.40	0.79	0.11
OSM24-48PSC Localized Family							
Weights	0.68	0.70	0.32	0.73	0.44	0.77	0.35
Membership A.	0.70	0.70	0.33	0.71	0.44	-	-
1	0.73	0.52	0.14	0.25	0.18	0.32	0.50
2	0.88	0.85	0.80	0.80	0.80	0.89	0.79
3	0.66	0.68	0.46	0.70	0.56	0.72	0.46
4	0.60	0.69	0.25	0.82	0.38	0.74	0.38
5	0.74	0.72	0.31	0.69	0.43	0.78	0.53

3.2   Localized Models Proof of Concept

To show that in a dataset with greater separability, localized models can give better performances, the centroids of the previously obtained clusters were taken and the 75 closest observations for train and the 50 closest test observations were extracted. Three new models were created, NEW 3, NEW 4 and NEW 5. Cluster 1 and 2 models, because of the already low amount of data present, were used as before. For comparison reasons, a global model with new data with greater separability was created (NEW GLOBAL). The results are present in Table 5, with a new silhouette score between all clusters of 0.2, evidencing a larger separability than the one obtained before (0.11).

In Table 5, the cross-validation performance (except precision) and the test recall of the combined models is larger than the one from the global model, which may indicate that this approach is useful in more separable populations. The test recall is especially of interest because it shows that the combined models correctly identify more mortality observations. However, the F1 Score and precision of the combined new models is smaller. On the surface, this would not be expected, however, when taking into account the limited amount of data and the resulting scarcity of positive mortality observations in some clusters, this can be explained. These characteristics significantly limit the predictive capabilities, particularly impacting this type of analysis. Cluster 1 stands out as particularly problematic in this regard. This leads to conclude that localized models would probably be useful in more separable data, but not compromised by imbalanced class distribution and a small amount of data. Nonetheless, further research is still necessary to fully understand and validate this approach.

Table 5. Performance of the NEW 3, NEW 4 and NEW 5 models, the weighted average (NEW COMBINED) of the new models in conjunction with 1 and 2 of the M24-24PSC family and the NEW GLOBAL model.

Models		Accuracy	Balanced Acc.	Precision	Recall	F1-score
NEW	C. Val.	0.69 ± 0.13	0.67 ± 0.14	0.69 ± 0.26	0.53 ± 0.22	0.58 ± 0.18
3	Test P.	0.72	0.72	0.57	0.71	0.63
NEW	C. Val.	0.87 ± 0.11	0.63 ± 0.25	0.30 ± 0.46	0.30 ± 0.46	0.30 ± 0.46
4	Test P.	0.72	0.77	0.28	0.83	0.42
NEW	C. Val.	0.86 ± 0.09	0.59 ± 0.20	0.20 ± 0.40	0.20 ± 0.40	0.20 ± 0.40
5	Test P.	0.64	0.81	0.14	1.00	0.25
NEW	C. Val.	0.83 ± 0.11	0.67 ± 0.20	0.36 ± 0.29	0.33 ± 0.28	0.33 ± 0.27
COMBINED	Test P.	0.68	0.77	0.34	0.88	0.45
NEW	C. Val.	0.77 ± 0.11	0.60 ± 0.13	0.46 ± 0.41	0.30 ± 0.26	0.32 ± 0.27
GLOBAL	Test P.	0.72	0.76	0.39	0.82	0.53

4   Conclusions

This work had the objective of predicting the probability of mortality in patients with pneumonia after entrance into ICU. The usefulness of localized models was also accessed, to predict outcomes in different populations of the data.

The best models obtained were M24-48PS and Weights M24-48PSC, both with 0.73 of balanced accuracy and 0.49 and 0.48 in F1-score, respectively. Unfortunately, the F1-score is low, due to the low precision. The results of the global and combined localized models do not reveal major differences in performance, due to the low separability of the data, however, the proof of concept showed that it has the potential to be useful in more separable data. Nevertheless, the localized models achieved different performances separately and got different importance in the mortality indicators, showing that the model performance and mortality indicators depend on the population.

The obtained AUROC value of 0.78 for the best models is lower compared to most of the models presented in Table 1. This difference can be mainly attributed to the fact that in current literature, more complex models than the one used in this work are usually used, that permit to get better performances. However, the choice of more complex models could hinder the search for localized models and compromise the interpretability of the results, which is especially important in a medical context. The other reason that explains these results is the fact that very heterogeneous and varied data was used, which does not always happen in other models.

Future work will explore different approaches aiming to increase the performance of the predictive models, including optimizing data pre-processing parameters, as well as clustering analysis, and considering other clinical variables (or other codifications of the existing ones). It will also experiment with more complex models that permit better performances. Lastly, it would be important to continue this work together with professionals in the field of health, who can provide insights on how to tune the models according to their needs. Overall, it is very important that this research for optimizing the prediction of mortality in pneumonia is continued, in an effort to improve health care.

Acknowledgments. This work was partially funded by FCT, I.P. through national funds, within the scope of the UIDB/00127/2020 project (IEETA/UA, http://www.ieeta.pt).

References

1. Antunes, C., et al.: Hospitalization direct cost of adults with community-acquired pneumonia in Portugal from 2000 to 2009. Pulmonology **26**(5), 264–267 (2020). https://doi.org/10.1016/J.PULMOE.2020.02.013, https://www.journalpulmonology.org/en-hospitalization-direct-cost-adults-with-articulo-S2531043720300969
2. Casillas, N., Torres, A.M., Moret, M., Gómez, A., Rius-Peris, J.M., Mateo, J.: Mortality predictors in patients with COVID-19 pneumonia: a machine learning approach using eXtreme gradient boosting model. Internal Emerg. Med. **17**(7), 1929–1939 (2022). https://doi.org/10.1007/S11739-022-03033-6/TABLES/3, https://link.springer.com/article/10.1007/s11739-022-03033-6
3. Chen, Y.M., et al.: Real-time interactive artificial intelligence of things-based prediction for adverse outcomes in adult patients with pneumonia in the emergency department. Acad. Emerg. Med. **28**(11), 1277–1285 (2021). https://doi.org/10.1111/acem.14339, https://onlinelibrary.wiley.com/doi/full/10.1111/acem.14339
4. Cilloniz, C., et al.: Machine-learning model for mortality prediction in patients with community-acquired pneumonia. Chest (2022). https://doi.org/10.1016/j.chest.2022.07.005, https://pubmed.ncbi.nlm.nih.gov/35850287/
5. Cox, D.R.: The regression analysis of binary sequences. J. Royal Statist. Soc.. Ser. B (Methodological) **20**(2), 215–242 (1958). http://www.jstor.org/stable/2983890
6. Dormann, C.F., et al.: Collinearity: A review of methods to deal with it and a simulation study evaluating their performance. Ecography **36**(1), 27–46 (2013). https://doi.org/10.1111/j.1600-0587.2012.07348.x, https://onlinelibrary.wiley.com/doi/full/10.1111/j.1600-0587.2012.07348.x
7. Fine, M.J., et al.: A prediction rule to identify low-risk patients with community-acquired pneumonia. New England J. Med. **336**(4), 834 (1997). https://doi.org/10.1056/NEJM199701233360402, https://pubmed.ncbi.nlm.nih.gov/8995086/
8. Harris, C.R., et al.: Array programming with NumPy. Nature **585**(7825), 357–362 (2020). https://doi.org/10.1038/s41586-020-2649-2
9. He, H., Bai, Y., Garcia, E.A., Li, S.: ADASYN: adaptive synthetic sampling approach for imbalanced learning. In: Proceedings of the International Joint Conference on Neural Networks, pp. 1322–1328 (2008). https://doi.org/10.1109/IJCNN.2008.4633969

10. IHME: Global Burden of Disease (GBD 2019) | Institute for Health Metrics and Evaluation (2019). https://www.healthdata.org/gbd/2019

11. Le Gall, J.R., Lemeshow, S., Saulnier, F.: A new simplified acute physiology score (SAPS II) based on a European/North American multicenter study. JAMA **270**(24), 2957–2963 (1993). https://doi.org/10.1001/JAMA.270.24.2957, https://pubmed.ncbi.nlm.nih.gov/8254858/

12. Kang, S.Y., et al.: Predicting 30-day mortality of patients with pneumonia in an emergency department setting using machine-learning models. Clin. Exper. Emerg. Med. **7**(3), 197–205 (sep 2020). https://doi.org/10.15441/ceem.19.052, https://www.ncbi.nlm.nih.gov/pmc/articles/PMC7550804/

13. Knaus, W.A., Draper, E.A., Wagner, D.P., Zimmerman, J.E.: APACHE II: a severity of disease classification system. Crit. Care Med. **13**(10) (1985). https://doi.org/10.1097/00003246-198510000-00009

14. Laino, M.E., et al.: An individualized algorithm to predict mortality in COVID-19 pneumonia: a machine learning based study. Arch. Med. Sci. **18**(3), 587–595 (2022). https://doi.org/10.5114/AOMS/144980, https://www.archivesofmedicalscience.com/An-individualized-algorithm-to-predict-mortality-in-COVID-19-pneumonia-a-machine,144980,0,2.html

15. Lee, J., Dubin, J.A., Maslove, D.M.: Mortality prediction in the ICU. In: Secondary Analysis of Electronic Health Records, pp. 315–324. Springer, Cham (2016). https://doi.org/10.1007/978-3-319-43742-2_21

16. Leiner, J., et al.: Machine learning-derived prediction of in-hospital mortality in patients with severe acute respiratory infection: analysis of claims data from the German-wide Helios hospital network. Respir. Res. **23**(1), 1–12 (2022). https://doi.org/10.1186/S12931-022-02180-W/FIGURES/3, https://respiratory-research.biomedcentral.com/articles/10.1186/s12931-022-02180-w

17. Lemaître, G., Nogueira, F., Aridas, C.K.: Imbalanced-learn: a python toolbox to tackle the curse of imbalanced datasets in machine learning. J. Mach. Learn. Res. **18**(17), 1–5 (2017). http://jmlr.org/papers/v18/16-365.html

18. Lim, W.S., et al.: Defining community acquired pneumonia severity on presentation to hospital: an international derivation and validation study. Thorax **58**, 377–382 (2003). https://doi.org/10.1136/thorax.58.5.377, www.thoraxjnl.com

19. Mandell, L.A., et al.: Infectious Diseases Society of America/American Thoracic Society consensus guidelines on the management of community-acquired pneumonia in adults. Clinical infectious diseases: an official publication of the Infectious Diseases Society of America 44 Suppl 2(Suppl 2) (2007). https://doi.org/10.1086/511159, https://pubmed.ncbi.nlm.nih.gov/17278083/

20. McKinney, W.: Data structures for statistical computing in python. In: van der Walt, S., Millman, J. (eds.) Proceedings of the 9th Python in Science Conference, pp. 51–56 (2010)

21. McLuckie, A.: Respiratory Disease and Its Management. Springer, London (2009). https://doi.org/10.1007/978-1-84882-095-1

22. Pedregosa, F., et al.: Scikit-learn: machine learning in Python. J. Mach. Learn. Res. **12**, 2825–2830 (2011)

23. Pirracchio, R., Petersen, M.L., Carone, M., Rigon, M.R., Chevret, S., van der Laan, M.J.: Mortality prediction in intensive care units with the Super ICU Learner Algorithm (SICULA): a population-based study. Lancet Respir. Med. **3**(1), 42–52 (2015). https://doi.org/10.1016/S2213-2600(14)70239-5

24. Regunath, H., Oba, Y.: Community-Acquired Pneumonia. StatPearls (2022). https://www.ncbi.nlm.nih.gov/books/NBK430749/

25. Rousseeuw, P.J.: Silhouettes: a graphical aid to the interpretation and validation of cluster analysis. J. Comput. Appl. Math. **20**(C), 53–65 (1987). https://doi.org/10.1016/0377-0427(87)90125-7

26. Ryan, L., et al.: Mortality prediction model for the triage of COVID-19, pneumonia, and mechanically ventilated ICU patients: a retrospective study. Ann. Med. Surg. **59**, 207–216 (2020). https://doi.org/10.1016/J.AMSU.2020.09.044

27. Sidey-Gibbons, J.A., Sidey-Gibbons, C.J.: Machine learning in medicine: a practical introduction. BMC Med. Res. Methodol. **19**(1), 1–18 (2019). https://doi.org/10.1186/S12874-019-0681-4/TABLES/5, https://link.springer.com/articles/10.1186/s12874-019-0681-4

28. Tong, S., Amand, C., Kieffer, A., Kyaw, M.H.: Trends in healthcare utilization and costs associated with pneumonia in the United States during 2008–2014 11 Medical and Health Sciences 1117 Public Health and Health Services. BMC Health Serv. Res. **18**(1), 1–8 (2018). https://doi.org/10.1186/S12913-018-3529-4/TABLES/6, https://bmchealthservres.biomedcentral.com/articles/10.1186/s12913-018-3529-4

29. Vincent, J.L., et al.: The SOFA (Sepsis-related Organ Failure Assessment) score to describe organ dysfunction/failure. On behalf of the working group on sepsis-related problems of the European society of intensive care medicine. Intensive Care Med. **22**(7), 707–710 (1996). https://doi.org/10.1007/BF01709751, https://pubmed.ncbi.nlm.nih.gov/8844239/

30. Virtanen, P., et al.: SciPy 1.0: fundamental algorithms for scientific computing in Python. Nat. Methods **17**, 261–272 (2020). https://doi.org/10.1038/s41592-019-0686-2

31. Ward, J.H.: Hierarchical grouping to optimize an objective function. J. Am. Stat. Assoc. **58**(301), 236–244 (1963). https://doi.org/10.1080/01621459.1963.10500845

32. Welte, T., Torres, A., Nathwani, D.: Clinical and economic burden of community-acquired pneumonia among adults in Europe. Thorax **67**(1), 71–79 (2012). https://doi.org/10.1136/THX.2009.129502, https://pubmed.ncbi.nlm.nih.gov/20729232/

33. Yeo, I.K., Johnson, R.A.: A new family of power transformations to improve normality or symmetry. Biometrika **87**(4), 954–959 (2000). http://www.jstor.org/stable/2673623

ECG Feature-Based Classification of Induced Pain Levels

Daniela Pais[1]($\boxtimes$) and Raquel Sebastião[1,2]

[1] IEETA, DETI, LASI, University of Aveiro, 3810-193 Aveiro, Portugal
{danielapais,raquel.sebastiao}@ua.pt
[2] Polytechnic of Viseu, 3504-510 Viseu, Portugal

Abstract. Appropriate pain treatment relies on an accurate assessment of pain. Limitations regarding subjective reporting of pain or observational bias, when pain is assessed by a healthcare professional, can lead to inadequate pain treatment. Therefore, pain assessment using physiological signals has been studied in past years due to the importance of objective measurement. The aim of this work is to use features extracted from Electrocardiogram (ECG) signals to classify pain induced by a Cold Pressor Task (CPT). Specifically, the goal is to determine the optimal hyperparameters of the classification algorithms and the optimal features for accurately distinguishing between higher and lower levels of pain. A model combining 15 ECG-features related to the P, R, S, and T waves and the Random Forest algorithm provided the best performance for predicting induced pain levels. This model achieved an accuracy of 95.3%, an F1-score of 94.0%, a precision of 97.9%, and a recall of 90.4%. These results show the feasibility of identifying pain through the physiological characteristics of the ECG.

Keywords: Cold Pressor Task (CPT) · Classification · Electrocardiogram (ECG) · Induced pain · Machine learning · Pain assessment

1 Introduction and Background

An accurate assessment of pain intensity is crucial for effective pain management [10]. Currently, pain is typically assessed through self-reporting using scales and questionnaires, both in clinical and experimental settings. For instance, one widely used approach is the Numerical Rating Scale (NRS), where patients rate their pain on a scale from 0 to 10 (or 100), representing the absence of pain to the worst imaginable pain [5]. While self-reporting is considered the most appropriate method for accurately assessing pain, it may not always be feasible, particularly in individuals with limited verbal communication ability or cognitive impairment [9]. In clinical settings where patients are unable to self-report their pain, healthcare providers assess pain intensity by considering physiological indicators and behavioral cues to derive a final pain score. Nonetheless, this assessment process may be susceptible to the influence of observational bias [12].

© Springer Nature Switzerland AG 2024
V. Vasconcelos et al. (Eds.): CIARP 2023, LNCS 14470, pp. 45–59, 2024.
https://doi.org/10.1007/978-3-031-49249-5_4

Artificial Intelligence (AI) has shown advantages in accurately assessing pain and recent research has been exploring physiological signals to support it [17]. The effects of the Autonomic Nervous System (ANS) in response to pain can be measured non-invasively through physiological signals, allowing for the detection of increased sympathetic activity related to pain through physiological changes rather than relying on self-report [10]. The work [6] showed that k-Nearest Neighbor (KNN), Linear Discriminant Analysis (LDA), and Support Vector Machine (SVM) classifiers could identify pain, induced by external electrical stimulation, with accuracies of 83.94%, 84.28%, and 96.47%, respectively, using Skin Conductance Level (SCL), Blood Volume Pulse (BVP), and Electrocardiogram (ECG) signals. In [13], an SVM classifier was used to classify pain levels based on ECG signals, achieving validation accuracies of 62.72% for the classification between the baseline and the highest pain levels category, and 84.14% for the classification between the baseline and moderate pain levels category. The work [2] achieved an accuracy of 82.8% for the classification of pain using Electromyography (EMG), SCL, and ECG signals. The approach involved data augmentation and feature selection prior to training an SVM-rbf (SVM with Radial Basis Function kernel) model. In [11], a pain classifier based on a Deep Belief Network (DBN) using Photoplethysmography (PPG) was proposed. This method achieved an accuracy of 86.79% in distinguishing between the absence of pain during the preoperative period and the presence of pain in the immediate postoperative period. Furthermore, the DBN-based classifier achieved an accuracy of 65.57% in a multi-class classification approach for distinguishing between the absence of pain, and mild, moderate, and severe pain. In the work [17], a classification method based on deep neural networks with a late fusion approach using ECG, EMG, and Electrodermal Activity (EDA) signals achieved an accuracy of 84.40% for the classification between baseline and pain using the BioVid Heat Pain Database. However, this approach demonstrated an accuracy of 84.57% using only the EDA signal.

The aim of this study was to investigate the responses of ECG signals to pain through a controlled pain-inducing procedure involving thermal stimulation, the Cold Pressor Task (CPT) [16], as a step towards developing an AI physiological method for objective pain assessment. This method aims to assist healthcare providers in assessing, monitoring, and treating pain in clinical practice, contributing to better patient care. The ECG records the electrical activity of the heart and represents a sequence of cardiac cycles, which is composed of the P wave, the QRS complex, and the T wave [15]. This paper analyzes features based on ECG waves, with the goal of determining the most relevant ones for classifying pain. This work aims to perform model and algorithm comparisons to identify the most appropriate model for accurately distinguishing between two pain levels, low/moderate pain and high pain.

The paper is organized as follows: Sect. 2 presents the protocol implemented for data collection. In addition, this section outlines the proposed methodology for data preprocessing and processing and provides a description of the extracted features and the Machine Learning (ML) methods employed for the classification of pain induced during CPT. In Sects. 3 and 4, the findings are presented and discussed, respectively. The last section contains the conclusions and future work.

2   Methods

This section describes the experimental protocol for data collection and explains the methods implemented for analyzing the ECG responses during pain induction through cold pain stimuli applied as a CPT.

2.1   Dataset

The dataset comprises 642 examples and consists of data from 37 participants, 23 female and 14 male, with ages ranging from 19 to 25 years old (21.36 ± 1.27 years old). All volunteers were recruited from the university student community.

2.2   Experimental Protocol and Data Collection

A representation of the stages of the experimental procedure of the data collection protocol is shown in Fig. 1. This study was approved by the Ethics and Deontological Council of the University of Aveiro (CED-UA-24-CED/2021).

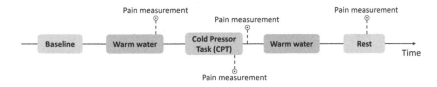

Fig. 1. Experimental procedure for data collection.

During the initial phase of the protocol, the participant was asked to sit in a comfortable position, while a five-minute baseline was recorded. Following the baseline recording, participants were instructed to immerse their nondominant hand and forearm in a warm water tank for two minutes to ensure a consistent skin temperature across the participants before the CPT. After, the participants submerged their nondominant forearm in a cold water tank with a temperature of approximately $7\,^{\circ}C\pm1\,^{\circ}C$. Participants were asked to endure the pain for as long as they could, with a time limit of two minutes. If they could not tolerate the pain, they were encouraged to inform the researcher and, before withdrawing their arm, to report their pain level. If they were able to complete the CPT, they were asked to report their maximum discomfort around the two-minute mark. Participants were required to report their pain level using the NRS. Afterward, participants were instructed to immerse their nondominant hand and forearm in the warm water tank for another two minutes of immersion. Before the end of the protocol, the participants were at rest while sitting in a comfortable position for five minutes. The ECG was recorded continuously using minimally invasive equipment during the entire protocol.

More details regarding the experimental procedure for data collection, including inclusion and exclusion criteria, ethical considerations, and the data collection setup can be consulted in the work [16].

2.3   Methods for Dataset Analysis

The goal of this study was to analyze the physiological changes in the ECG signal caused by pain. Therefore, for this analysis, only the data collected during the CPT, which regards the pain induction phase, was considered.

This work performed binary classification between low and medium pain (NRS score < 8) and high pain (NRS score ≥ 8), with the goal of identifying the most relevant ECG features and optimal hyperparameters of the classification algorithms for accurately distinguishing these pain categories. The experiments were performed in Python, mainly using scikit-learn (which supports supervised learning).

Feature Extraction and Normalization. The dataset includes 21 features extracted from the ECG signals in 20-second periods with a 75% overlap. The features, described in Table 1, were computed based on the location of the peaks of the P, R, S, and T waves and the onsets and offsets from the P, R, and T waves (see Fig. 2). The extracted features were normalized by dividing each epoch by the average of the respective feature in the baseline (see Fig. 1).

Table 1. Description of the extracted ECG features.

ECG Feature	Description
P,R,S,T_amplitude	Average amplitude of P, R, S and T waves
P,R,S,T_distance	Average distance between each corresponding wave
P,R,S,T_peaks	Number of peaks of P, R, S and T waves
P,R,T_onsetamp	Amplitude of the onset of P, R, and T waves
P,R,T_offsetamp	Amplitude of the offset of P, R, and T waves
P,R,T_onoffdist	Average distance between the onset and offset of P, R, and T waves

Fig. 2. Location of the extracted peaks, onsets and offsets of an ECG cycle.

Feature Selection. Feature selection was accomplished to remove redundant features and reduce the computational complexity and execution time. Two methods were compared:

1. Filter method: features with a correlation coefficient higher than 0.9 were considered strongly correlated, and the one with lower variance was removed from each pair of highly correlated features. The correlation between the features was evaluated using the Spearman correlation coefficient, as the features did not follow a normal distribution.
2. Wrapper method: sequential feature selection through a backward selection was used to select sets composed of 2 to 20 features. Backward elimination starts with the set of all available features, and one feature at a time is removed.

Feature Scaling. Standardization of the features was employed for the classification models which are based on distance measures to guarantee that variations in the magnitude and range of the features do not influence the results of the classification task, thus ensuring a fair and equitable comparison.

Classification. Binary classification was performed to distinguish between low and moderate pain levels (NRS score < 8) and high pain levels (NRS score $\geq$ 8). For comparing the classification algorithms, nested cross-validation (CV) was used. It is a recommended method for comparing algorithms in small to moderate-sized datasets. The nested CV involves two nested k-fold CV loops. The inner loop (2-fold CV in the inner loop) is used for model selection and the outer loop (5-fold CV in the outer loop) provides an estimate of the generalization performance. This approach has been shown to reduce bias in hyperparameter tuning and evaluation compared to traditional k-fold cross-validation, as reported in the study by Varma and Simon [18].

The original dataset was divided into train and test data (80/20) in a stratified fashion in order to maintain the original class proportion in the resulting subsets. The test data was set aside for the final evaluation of the model selected through nested CV on the train data. Since the performance estimates may suffer from pessimist bias if the training set is too small, the data from the outer loop was merged and used to fit the best model after model selection using nested CV. Finally, the generalization performance of the models was evaluated using the independent test set.

Six ML algorithms were evaluated, namely kNN, SVM, decision tree (DT), random forest (RF), adaptive boosting (AdaB), and extreme gradient boosting (XGB). The kNN approach is based on computing distances and performing classification predictions based on the majority vote of its nearest examples [8]. SVMs construct a hyperplane as the decision boundary between the two classes [1]. DTs learn simple decision rules based on the data features and implement recursive partitioning to construct the tree nodes [3]. RF is an ensemble method that builds on a set of several individual DTs, trained in parallel, which all contribute, equally, to make a final prediction [4]. AdaB and XGB are gradient-boosting algorithms, training several models in sequence emphasizing the training samples according to previous misclassification. These ensemble models typically use DTs as base learners, combining the performance of a set of weak learners to create a single strong learner [7,14]. Table 2 lists the hyperparameters that were tuned for each algorithm. The optimal hyperparameters of the classification algorithms were searched using scikit-learn `GridSearchCV` function by maximizing the F1-score, which was calculated by averaging the CV F1-score of the two inner loop splits. In addition to the F1-score, accuracy, precision, and recall scores were also used to assess the performance of each model. These metrics were also calculated in the outer loop for selecting the best models according to the generalization performance.

Table 2. Tuned hyperparameters of the classification algorithms.

Algorithm	Hyperparameters
kNN	p_neighbors: [1, 2, 3, 4, 5, 6, 7, 8, 9], p: [1, 2]
SVM	kernel: 'rbf', C: [0.1, 1, 10, 100, 1000], gamma: [0.00001, 0.0001, 0.001, 0.01, 0.1]
DT	criterion: ['gini', 'entropy'], max_depth: [5, 10, 15, 20, 25, 30, 35, 40, 45, 'None']
RF	criterion: ['gini', 'entropy'], max_depth: [5, 10, 15, 20, 25, 30, 35, 40, 45, 'None'] n_estimators: [10, 50, 100, 500]
AdaB	n_estimators: [10, 50, 100, 500], learning_rate: [0.01, 0.1, 0.5, 1]
XGB	n_estimators: [10, 50, 100, 500], learning_rate: [0.01, 0.1, 0.5, 1], max_depth: [2, 4, 6, 8]

Feature Importance. The importance of the features was computed for each model. For kNN and SVM algorithms, the feature importance was determined through the evaluation of feature permutation. Regarding DT, RF, AdaB and XGB, the importance was obtained through the total reduction of the criterion used to choose the best split at each node.

3   Results

The participants were required to report their pain level before the end of the CPT. The average value reported was 7.29 ± 1.58 (mean $\pm$ standard deviation), while the distribution of the pain levels had a median value of 7.

To conduct binary classification between two levels of pain intensity, the samples were divided into two groups based on the pain score distribution. Samples with pain scores greater than the median value (NRS score ≥ 8), which correspond to high pain, were assigned to the positive class, with 259 examples. The negative class, consisting of low and moderate pain levels (NRS score < 8), has 383 examples.

3.1 Datasets

The dataset was split into a training dataset and a test dataset. The training dataset consists of 513 samples, with 106 negative samples and 207 positive samples. The test dataset, on the other hand, consists of 129 samples, with 77 negative samples and 52 positive samples. The ratio between the positive and negative classes in both the train and test datasets is approximately 0.68, indicating a reasonably balanced dataset.

3.2 Classification

Table 3 presents the performance metrics of the models employing the original set of 21 ECG features listed in Table 1. These results correspond to the performance of the ML models using the test dataset.

Table 3. Performance evaluation metrics of the models using the test dataset, utilizing all 21 ECG features (the highest result for each performance metric is identified in bold). The optimal hyperparameters of each algorithm are also listed.

Model	Accuracy	F1-score	Precision	Recall
kNN (n_neighbors = 1, p = 1)	0.930	0.913	0.922	0.904
SVM (C = 10.0, gamma = 0.1, kernel = 'rbf')	0.876	0.855	0.810	0.904
DT (criterion = 'entropy', max_depth = 10)	0.876	0.857	0.800	**0.923**
RF (criterion = 'entropy', max_depth = 15, n_estimators = 500)	0.946	0.923	**0.979**	0.885
AdaB (learning_rate = 1, n_estimators = 100)	0.922	0.904	0.904	0.904
XGB (learning_rate = 0.1, max_depth = 4, n_estimators = 100)	**0.953**	**0.940**	**0.979**	0.904

Table 4 displays the test performance of the models using the features selected through the filter method based on the pairwise correlation of the features. The filter method was applied to the 21 features, removing six features (R_distance, S_distance, T_distance, Rpeaks, Speaks, and Tpeaks). Therefore, the resulting dataset contains 15 features (P,R,S,T_amplitude, P_distance, P_peaks, P,R,T_onsetamp, P,R,T_offsetamp, P,R,T_onoffdist) for classification.

Table 5 shows the best results for each algorithm obtained using the wrapper feature selection method previous to nested CV. The classification models with the highest performance were obtained by combining the optimal hyperparameters of each algorithm with the optimal features selected with backward feature selection. In all models, eleven features were consistently used from the original set, including P_distance, P_onoffdist, P_offsetamp, R_amplitude, R_onoffdist, R_onsetamp, S_amplitude, T_amplitude, T_onoffdist, T_onsetamp, and

Table 4. Performance evaluation metrics of the models using the test dataset with the features selected through the filter method (the highest result for each performance metric is identified in bold). The optimal hyperparameters of each algorithm are also listed.

Model	Accuracy	F1-score	Precision	Recall
kNN (n_neighbors = 1, p = 1)	0.946	0.931	0.959	0.904
SVM (C = 10, gamma = 0.1, kernel = 'rbf')	0.860	0.833	0.804	0.865
DT (criterion = 'entropy', max_depth = 10)	0.891	0.870	0.839	0.904
RF (criterion = 'entropy', max_depth = 10, n_estimators = 50)	**0.953**	**0.940**	**0.979**	0.904
AdaB (learning_rate = 1, n_estimators = 500)	0.946	0.923	**0.979**	0.885
XGB (learning_rate = 0.1, max_depth = 8, n_estimator s = 100)	0.946	0.932	0.941	**0.923**

T_offsetamp. Most of these features concern the onsets and offsets of the P, R, and T waves.

The test results, displayed in Tables 3, 4, and 5, for distinguishing between low and moderate (NRS score < 8) and high (NRS score ≥ 8) pain categories are summarized in Fig. 3.

Figures 4 and 5 display the scatter plots of the two principal components of the samples, highlighting the misclassified instances, which are categorized as false positives and false negatives for each model. Figure 4 shows the scatter plots depicting the results of the models using the set of 21 features. Figure 5

Table 5. Performance evaluation metrics of the models using the test dataset with the features selected through the wrapper method (the highest result for each performance metric is identified in bold). The optimal hyperparameters of each algorithm and the number of optimal features are also listed.

Model	Number of features	Accuracy	F1-score	Precision	Recall
kNN (n_neighbors = 1, p = 2)	11	0.884	0.857	0.849	0.865
SVM (C = 10, gamma = 0.1, kernel = 'rbf')	12	0.907	0.887	0.870	0.904
DT (criterion = 'entropy', max_depth = 15)	20	0.876	0.857	0.800	0.923
RF (criterion = 'gini', max_depth = 10, n_estimators = 500)	18	0.938	0.920	0.958	0.885
AdaB (learning_rate = 1, n_estimators = 500)	12	**0.946**	**0.931**	**0.959**	0.904
XGB (learning_rate = 0.1, max_depth = 4, n_estimators = 100)	15	0.930	0.916	0.891	**0.942**

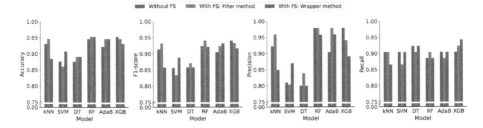

Fig. 3. Performance of the classification models using the test dataset for distinguishing between low and moderate (NRS score < 8) and high (NRS score ≥ 8) pain. FS: Feature selection.

displays the scatter plots of the models that achieved superior performances after employing feature selection, which are highlighted in Tables 4 and 5. The two principal components were obtained through principal component analysis (PCA) in order to reduce dimensionality by projecting the testing data to a 2D dimensional space, for data visualization purposes. Two scatter plots are provided for each model: one displaying all the samples in the plot, while the other provides a zoomed-in view of the plot. These scatter plots provide a visual representation of how well the classification models were able to distinguish between the two pain groups based on the set of ECG features used.

Figure 6 identifies the features included in the models whose classification results are illustrated in Fig. 5 and their corresponding feature importance in descending order. Particularly, both P_offsetamp and T_onsetamp demonstrate high importance across all four models.

4 Discussion

In general, the models were able to distinguish higher pain from low and moderate pain measured during the CPT with good performance (Fig. 3).

Globally, when training the models with the set of 21 features (Table 3), SVM and DT models had the worst results. However, the DT model (criterion='entropy', max_depth=10) achieved the highest recall score of 92.3%. Therefore, this model presented the highest capability to correctly classify the higher pain samples. kNN, SVM, AdaB and XGB all achieved a recall score of 90.4%. Therefore, less than 10% of the high pain samples were misclassified as lower pain levels. XGB model (learning_rate=0.5, n_estimators=500) provided the best results, with an accuracy score of 95.3% and an F1-score of 94.0%. Moreover, XGB and RF models exhibited the highest precision (97.9%) among all the models, indicating a high capability in classifying the negative samples (lower pain levels).

In Fig. 4, the boundary between the two classes is not clearly defined for any of the models. When examining the classification results obtained using kNN, RF, and XGB models, it is observed that only samples that are located in a

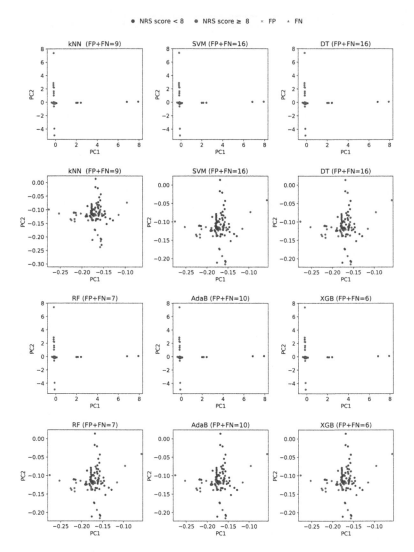

Fig. 4. Scatter plots representing the classification test results of the two groups of pain levels: low and moderate (NRS score < 8) and high (NRS score ≥ 8), using 21 ECG features. Each scatter plot is a 2D representation of the set of 21 features. The positive and negative class samples are shown in different colors, with the misclassified samples further distinguished between false positives (FP) and false negatives (FN).

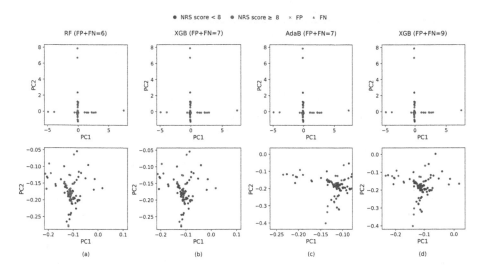

Fig. 5. Scatter plots representing the classification test results of the two groups of pain levels: low and moderate (NRS score < 8) and high (NRS score ≥ 8) for the: (a) RF and (b) XGB models using 15 ECG features selected using the filter method and (c) AdaB and (d) XGB models using 12 and 15 ECG features, respectively, selected using the wrapper method. The positive and negative class samples are shown in different colors, with the misclassified samples further distinguished between false positives (FP) and false negatives (FN).

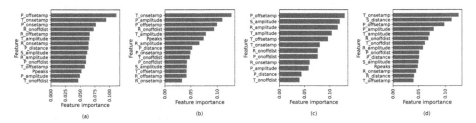

Fig. 6. Feature importance of the 15 features selected using the filter method employed with (a) RF and (b) XGB models and feature importance of the 12 and 15 features selected using the wrapper method implemented with (c) AdaB and (d) XGB models, respectively.

space with both negative and positive class samples in close proximity were misclassified, and one may reason that this could be justified by the lack of a clear boundary between both classes. However, when analyzing the results of SVM, DT, and AdaB models, it is observed that samples located at a greater distance from other observations were prone to misclassification, particularly negative samples (only the DT model presents one false negative).

Regarding the results obtained using the feature subset selected through pairwise feature correlation (Table 4), SVM and DT models also exhibited the lowest results. For this approach, the RF model (criterion='entropy', max_depth=10, n_estimators=50) had the best global performance, with an accuracy of 95.3% and an F1-score of 94.0%. The RF model performs well at identifying high-pain samples (recall=90.3%), but it may still incorrectly classify a small percentage of them as low and moderate pain levels, resulting in false negatives (Fig. 5 (a)) and the possibility of missing a participant in pain. Both RF and AdaB models had a high precision score of 97.9%, which indicates that the models are good at avoiding false positives. This precision score was also achieved in the previous approach with the RF and XGB algorithms, although with a larger feature set. In addition to the advantage of obtaining similar results to the previous approach with a smaller number of features, the RF model presented in Table 4 is less complex: an ensemble of 50 DT (10% of the ensemble of the previous approach) with a maximum depth of 10 (as opposed to 15). kNN also showed improved performance, with an accuracy of 94.6% and an F1-score of 93.1%. The kNN model has the advantage of a significantly lower mean fit time (0.002 s) compared to the RF model (0.892 s). The XGB model achieved a recall score of 92.3% (Fig. 5 (b)), using only 15 features, which matches the recall obtained with the DT model in the previous approach, despite the DT model employing a larger feature set consisting of 21 features.

Figure 6 shows that the feature importance attributed to the 15 features determined through pairwise correlation analysis of the original feature set exhibited variations between the RF (Fig. 6 (a)) and XGB (Fig. 6 (b)) models, resulting in distinct importance rankings for this set of features. Nonetheless, P_offsetamp, T_onsetamp, R_onoffdist, and T_amplitude were included in the six most significant features for both models.

Regarding the approach implementing backward feature selection (Table 5), the optimal number of features differed among the classification models. While DT and RF required a larger set of features, kNN, SVM, AdaB and XGB models performed well with a relatively smaller number of features. When compared to the previous results, the kNN model exhibited a significant decrease in performance (see Fig. 3). Similarly, the RF model demonstrated superior overall performance for the two previous approaches. SVM, on the other hand, improved performance. The AdaB model (learning_rate=1, n_estimators=500) was able to improve its performance as well. Using only 12 ECG features, the AdaB model led to an accuracy of 94.6% and an F1-score of 93.1%. Although reducing the number of features from 15 (selected using the filter method) to 12 features (selected using the wrapper method) only resulted in an improvement of 0.8% in the F1-score, the decrease in the number of features leads to reduced complexity of the model and faster run times. The AdaB model identified the amplitude (S,R,T_amplitude) and offset amplitude (P,T_offsetamp) of the ECG waves as the most significant features for the classification task (see Fig. 6 (c)). In contrast to the preceding two approaches, where a maximum precision score of 97.9% was achieved, the models resulting from the subset of features selected through sequential feature selection exhibited comparatively lower precision scores. Among these models, the AdaB model achieved the highest precision score of 95.9%, indicating that this model correctly identified 95.9% (FP=2) of the higher pain samples (Fig. 5 (c)). The XGB model (learning_rate=0.1, max_depth=4, n_estimators=100) achieved the highest recall score among all approaches with a score of 94.2%, using 15 features. This result indicates that 5.8% (3 samples) of the higher pain samples remained to be predicted, as displayed in Fig. 5 (d). XGB attributed the highest importance to features associated with various characteristics of the ECG wave. These features include the onset amplitude of the T wave (T_onsetamp), the offset amplitude of the P wave (P_offsetamp), the amplitude of the T and P waves (T_amplitude, P_amplitude), the distance between corresponding S waves (S_distance), and the distance between the onset and offset of the R waves (R_onoffdist). Despite having an equal number of features (n=15) to the subset selected through the filter method, the results achieved with this particular subset are inferior, further emphasizing the importance of selecting the most relevant and informative features. Regarding the subset selected through pairwise correlation analysis (Fig. 5 (b)), it achieved a precision of 94.1% (FP=3) and a recall of 92.3% (FN=4). In comparison, the subset selected using sequential feature selection (Fig. 5 (d)) achieved a precision of 89.1% (FP=6) and a recall of 94.2% (FN=3).

Finally, while the performance achieved by the models in all three approaches was similar, reducing the number of features results in a less complex model. Moreover, in the majority of the models, the mean fit time increased when using a larger feature set as opposed to a smaller one.

5   Conclusions and Further Research

The inadequate management of pain can result in psychological and physiological adverse effects. Both undertreatment and overtreatment may lead to complications. Therefore it is important to minimize false positives and false negatives in the detection of the level of pain in clinical settings. Thus, developing a model for pain management with high precision and high recall is crucial in order to prevent unnecessary treatment or undertreatment of patients and to ensure appropriate and effective management of pain.

This work evaluated a methodology for ECG feature-based classification of pain levels induced through a controlled pain-inducing procedure with cold thermal stimulation. The methodology involved optimizing the hyperparameters of the classification algorithms and identifying the most relevant features through feature selection for pain detection, resulting in models with reduced computational costs and time, that yielded comparable performances to the models presented in previous research of the authors.

The classification models were able to distinguish between higher and lower levels of pain measured during the CPT with high accuracy. The study showed that the ECG features related to the P, R, S, and T waves were effective in distinguishing between lower and higher pain levels. Overall, the models performed better at classifying lower pain samples than higher pain samples, as evidenced by the higher precision scores compared to the recall scores. However, XGB was particularly successful in correctly identifying higher pain samples, achieving a recall score of 94.2%, using only 15 ECG features. Among these features, T_onsetamp, S_distance, P_offsetamp, P_amplitude, and T_amplitude were identified in descending order of importance as the most relevant. Moreover, the RF algorithm, when combined with a different set of 15 ECG features, demonstrated the best overall performance in predicting pain levels, with an accuracy of 95.3%, an F1-score of 94.0%, a precision of 97.9%, and a recall of 90.4%. In decreasing order of importance, the most significant features were P_offsetamp, T_onsetamp, P_onsetamp, R_onoffdist, and R_offsetamp.

This work is an initial stage for an AI system that aims to support clinicians in objectively assessing pain for guiding drug administration. Although further research is required, the obtained results are a step further to contribute to an objective assessment of pain, which may lead to more personalized healthcare, ultimately improving the condition of patients. In future work, we propose investigating the ability of ECG to predict pain across more than two levels of pain through a multi-class classification approach. Additionally, exploring the use of deep learning is also suggested. Furthermore, integrating diverse physiological signals into a multi-signal assessment may improve the reliability and advance the development of more effective pain assessment methods.

Acknowledgements. This work was funded by national funds through FCT - Fundação para a Ciência e a Tecnologia, I.P., under the PhD grant UI/BD/153374/2022 (D.P.), under the Scientific Employment Stimulus CEECIND/03986/2018 (R.S.) and CEECINST/00013/2021 (R.S.), within the R&D unit IEETA/UA (UIDB/00127/2020), and under the project EMPA (2022.05005.PTDC).

References

1. Aggarwal, C.C.: Data Mining. Springer, Cham (2015). https://doi.org/10.1007/978-3-319-14142-8
2. Al-Qerem, A.: An efficient machine-learning model based on data augmentation for pain intensity recognition. Egyptian Inform. J. **21**, 241-257 (2020). (https://doi.org/10.1016/j.eij.2020.02.006), https://doi.org/10.3389/fnins.2017.00279
3. Breiman, L.: Classification and Regression Trees (1984). https://doi.org/10.1201/9781315139470
4. Breiman, L.: Random forests - random features, pp. 1–29 (1999)
5. Breivik, H., et al.: Assessment of pain. Br. J. Anaesth. **101**(1), 17–24 (2008). https://doi.org/10.1093/bja/aen103
6. Chu, Y., Zhao, X., Han, J., Su, Y.: Physiological signal-based method for measurement of pain intensity. Front. Neurosci. **11**(279) (2017). https://doi.org/10.3389/fnins.2017.00279
7. Friedman, J., Hastie, T., Tibshirani, R.: Additive logistic regression: a statistical view of boosting. Ann. Stat. **28**(2), 337–407 (2000). https://doi.org/10.1214/aos/1016120463
8. Guo, G., Wang, H., Bell, D., Bi, Y., Greer, K.: Knn model-based approach in classification, vol. 2888, pp. 986–996 (2003). https://doi.org/10.1007/978-3-540-39964-3_62

9. Hummel, P., van Dijk, M.: Pain assessment: current status and challenges. Semin. Fetal Neonatal. Med. **11**(4), 237–245 (2006)
10. Ledowski, T., Bromilow, J., Wu, J., Paech, M.J., Storm, H., Schug, S.A.: The assessment of postoperative pain by monitoring skin conductance: results of a prospective study. Anaesthesia **62**(10), 989–993 (2007)
11. Lim, H., Kim, B., Noh, G.J., Yoo, S.K.: A deep neural network-based pain classifier using a photoplethysmography signal. Sensors **19**(2) (2019). https://doi.org/10.3390%2Fs19020384
12. Maxwell, L.G., Fraga, M.V., Malavolta, C.P.: Assessment of pain in the newborn: an update. Clin. Perinatol. **46**(4), 693–707 (2019). https://doi.org/10.1016/j.clp.2019.08.005
13. Naeini, E.K., et al.: Pain recognition with electrocardiographic features in post-operative patients: method validation study. J. Med. Internet Res. **23**(5), e25079 (2021). https://doi.org/10.2196/25079
14. Natekin, A., Knoll, A.: Gradient boosting machines, a tutorial. Front. Neurorobot. **7** (2013). https://doi.org/10.3389/fnbot.2013.00021
15. Saladin, S.K.: Human Anatomy, 5 edn. McGraw-Hill Education (2017)
16. Silva, P., Sebastião, R.: Using the electrocardiogram for pain classification under emotional contexts. Sensors **23**(3), 1443 (2023). https://doi.org/10.3390/s2303144
17. Thiam, P., Bellmann, P., Kestler, H.A., Schwenker, F.: Exploring deep physiological models for nociceptive pain recognition. Sensors **19**(4503) (2019). https://doi.org/10.3390/s19204503
18. Varma, S., Simon, R.: Bias in error estimation when using cross-validation for model selection. BMC Bioinform. **7**(91), 3242–3249 (2006). https://doi.org/10.1186/1471-2105-7-91

Leveraging Model Fusion for Improved License Plate Recognition

Rayson Laroca[1](✉)(iD), Luiz A. Zanlorensi[1](iD), Valter Estevam[1,2](iD),
Rodrigo Minetto[3](iD), and David Menotti[1](iD)

[1] Federal University of Paraná, Curitiba, Brazil
rblsantos@inf.ufpr.br
[2] Federal Institute of Paraná, Irati, Brazil
[3] Federal University of Technology-Paraná, Curitiba, Brazil

Abstract. License Plate Recognition (LPR) plays a critical role in various applications, such as toll collection, parking management, and traffic law enforcement. Although LPR has witnessed significant advancements through the development of deep learning, there has been a noticeable lack of studies exploring the potential improvements in results by fusing the outputs from multiple recognition models. This research aims to fill this gap by investigating the combination of up to 12 different models using straightforward approaches, such as selecting the most confident prediction or employing majority vote-based strategies. Our experiments encompass a wide range of datasets, revealing substantial benefits of fusion approaches in both intra- and cross-dataset setups. Essentially, fusing multiple models reduces considerably the likelihood of obtaining subpar performance on a particular dataset/scenario. We also found that combining models based on their speed is an appealing approach. Specifically, for applications where the recognition task can tolerate some additional time, though not excessively, an effective strategy is to combine 4–6 models. These models may not be the most accurate individually, but their fusion strikes an optimal balance between accuracy and speed.

Keywords: License Plate Recognition · Model Fusion · Ensemble

1 Introduction

Automatic License Plate Recognition (ALPR) has garnered substantial interest in recent years due to its many practical applications, which include toll collection, parking management, border control, and road traffic monitoring [18, 22, 43].

In the deep learning era, ALPR systems customarily comprise two key components: license plate detection (LPD) and license plate recognition (LPR). LPD entails locating regions containing license plates (LPs) within an image, while

Supported by the Coordination for the Improvement of Higher Education Personnel (CAPES) (# 88881.516265/2020-01), and by the National Council for Scientific and Technological Development (CNPq) (# 309953/2019-7 and # 308879/2020-1).

V. Vasconcelos et al. (Eds.): CIARP 2023, LNCS 14470, pp. 60–75, 2024.
https://doi.org/10.1007/978-3-031-49249-5_5

LPR involves identifying the characters within these LPs. Recent research has predominantly concentrated on advancing LPR [27,30,47], given that widely adopted object detectors such as Faster-RCNN and YOLO have consistently delivered impressive results in LPD for some years now [14,21,48].

This study also focuses on LPR but provides a unique perspective compared to recent research. Although deep learning techniques have enabled significant advancements in this field over the past years, multiple studies have shown that different models exhibit varying levels of robustness across different datasets [19, 29,46]. Each dataset poses distinct challenges, such as diverse LP layouts and varying tilt ranges. As a result, a method that performs optimally on one dataset may yield poor results on another. This raises an important question: *"Can we substantially enhance LPR results by fusing the outputs of diverse recognition models?"* If so, two additional questions arise: *"To what extent can this improvement be attained?"* and *"How many and which models should be employed?"* As of now, such questions remain unanswered in the existing literature.

We acknowledge that some ALPR applications impose stringent time constraints on their execution. This is particularly true for embedded systems engaged in tasks such as access control and parking management in high-traffic areas. However, in other contexts, such as systems used for issuing traffic tickets and conducting forensic investigations, there is often a preference to prioritize the recognition rate, even if it sacrifices efficiency [16,30,33]. These scenarios can greatly benefit from the fusion of multiple recognition models.

Taking this into account, in this study, we thoroughly examine the potential of enhancing LPR results through the fusion of outputs from multiple recognition models. Remarkably, we assess the combination of up to 12 well-known models across 12 different datasets, setting our investigation apart from earlier studies.

In summary, this paper has two main contributions:

- We present empirical evidence showcasing the benefits offered by fusion approaches in both intra- and cross-dataset setups. In the intra-dataset setup, the mean recognition rate across the datasets experiences a substantial boost, rising from 92.4% achieved by the best model individually to 97.6% when leveraging the best fusion approach. Similarly, in the cross-dataset setup, the mean recognition rate increases from 87.6% to levels exceeding 90%. Notably, in both setups, the sequence-level majority vote fusion approach outperform both character-level majority vote and selecting the prediction made with the highest confidence approaches.
- We draw attention to the effectiveness of fusing models based on their speed. This approach is particularly useful for applications where the recognition task can accommodate a moderate increase in processing time. In such cases, the recommended strategy is to combine 4–6 fast models. Although these models may not achieve the highest accuracy individually, their fusion results in an optimal trade-off between accuracy and speed.

The rest of this paper is structured as follows. Section 2 provides a concise overview of the recognition models explored in this work. The experimental setup adopted in our research is detailed in Sect. 3. The results obtained are presented and analyzed in Sect. 4. Lastly, Sect. 5 summarizes our findings.

2   Related Work

LPR is widely recognized as a specific application within the field of scene text recognition [7,26,50]. LPR sets itself apart primarily due to the limited presence of strong linguistic context information and the minimal variation observed between characters. The following paragraphs briefly describe well-known models originally proposed for general scene text recognition, LPR, and related tasks. These models will be explored in this study.

Baek et al. [2] introduced a four-stage framework (depicted in Fig. 1) that models the design patterns of most modern methods for scene text recognition. The *Transformation* stage removes the distortion from the input image so that the text is horizontal or normalized. This task is generally done through spatial transformer networks with a thin-plate splines (TPS) transformation, which models the distortion by finding and correcting fiducial points. The second stage, *Feature Extraction*, maps the input image to a representation that focuses on the attributes relevant to character recognition while suppressing irrelevant features such as font, color, size and background. This task is usually performed by a module composed of Convolutional Neural Networks (CNNs), such as VGG, ResNet, and RCNN. The *Sequence Modeling* stage converts visual features to contextual features that capture the context in the sequence of characters. Bidirectional Long Short-Term Memory (Bi-LSTM) is generally employed for this task. Finally, the *Prediction* stage produces the character sequence from the identified features. This task is typically done by a Connectionist Temporal Classification (CTC) decoder or through an attention mechanism. As can be seen in Table 1, while most methods can fit within this framework, they do not necessarily incorporate all four modules.

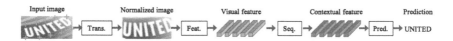

Fig. 1. The four modules or stages of modern scene text recognition, according to [2]. "Trans." stands for Transformation, "Feat." stands for Feature Extraction, "Seq." stands for Sequence Modeling, and "Pred." stands for Prediction. Image reproduced from [2].

Table 1. Summary of seven well-known models for scene text recognition. These models are listed chronologically and are further explored in other sections of this work.

Model	Transformation	Feature Extraction	Sequence Modeling	Prediction
R^{2}AM [25]	–	RCNN	–	Attention
RARE [35]	TPS	VGG	Bi-LSTM	Attention
STAR-Net [28]	TPS	ResNet	Bi-LSTM	CTC
CRNN [34]	–	VGG	Bi-LSTM	CTC
GRCNN [42]	–	RCNN	Bi-LSTM	CTC
Rosetta [4]	–	ResNet	–	CTC
TRBA [2]	TPS	ResNet	Bi-LSTM	Attention

Atienza [1] drew inspiration from the accomplishments of the Vision Transformer (ViT) and put forward a single-stage model named ViTSTR for scene text recognition. It operates by initially dividing the input image into non-overlapping patches. These patches are then converted into 1–D vector embeddings (i.e., flattened 2–D patches). To feed the encoder, each embedding is supplemented with a learnable patch embedding and a corresponding position encoding.

Recent works on LPR have focused on developing multi-task CNNs that can process the entire LP image holistically, eliminating the need for character segmentation [6,11,39]. Two such models are Holistic-CNN [39] and Multi-task-LR [11]. In these models, the LP image undergoes initial processing via convolutional layers, followed by N branches of fully connected layers. Each branch is responsible for predicting a single character class (including a 'blank' character) at a specific position on the LP, enabling the branches to collectively predict up to N characters. Both models are often used as baselines due to their remarkable balance between speed and accuracy [12,19,20,27,30].

The great speed/accuracy trade-off provided by YOLO networks [41] has inspired many authors to explore similar architectures targeting real-time performance for LPR and similar tasks. Silva & Jung [37] proposed CR-NET, a YOLO-based model that effectively detects and recognizes all characters within a cropped LP [19,22,38]. Another noteworthy model is Fast-OCR [23], which incorporates features from several object detectors that prioritize the trade-off between speed and accuracy. In the domain of automatic meter reading [23], Fast-OCR achieved considerably better results than multiple baselines that perform recognition holistically, including CRNN [34], Multi-task-LR [11] and TRBA [2].

While we found a few works leveraging model fusion to improve LPR results, we observed that they explored a limited range of models and datasets in the experiments. For example, Izidio et al. [16] employed multiple instances of the same model (i.e., Tiny-YOLOv3) rather than different models with varying architectures. Their experiments were conducted exclusively on a private dataset. Another example is the very recent work by Schirrmacher et al. [33], where they examined deep ensembles, BatchEnsemble, and Monte Carlo dropout using multiple instances of two backbone architectures. The authors' primary focus was on recognizing severely degraded images, leading them to perform nearly all of their experiments on a synthetic dataset containing artificially degraded images.

In summary, although the field of LPR has witnessed significant advancements through the development and application of deep learning-based models, there has been a noticeable lack of studies thoroughly examining the potential improvements in results by fusing the outputs from multiple recognition models.

3   Experimental Setup

This section provides an overview of the setup adopted in our experiments. We first enumerate the recognition models implemented for this study, providing specific information about the framework used for training and testing each of them, as well as the corresponding hyperparameters. Subsequently, we compile a list of the datasets employed in our assessments, showcasing sample LP images from each dataset to highlight their diversity. Afterward, we elaborate on the strategies examined for fusing the outputs of the different models. Finally, we describe how the performance evaluation is carried out.

The experiments were conducted on a PC with an AMD Ryzen Threadripper 1920X 3.5GHz CPU, 96 GB of RAM operating at 2133 MHz, an SSD (read: 535 MB/s; write: 445 MB/s), and an NVIDIA Quadro RTX 8000 GPU (48 GB).

3.1   Recognition Models

We explore 12 recognition models in our experiments: RARE [35], R^2AM [25], STAR-Net [28], CRNN [34], GRCNN [42], Holistic-CNN [39], Multi-task-LR [11], Rosetta [4], TRBA [2], CR-NET [37], Fast-OCR [23] and ViTSTR-Base [1]. As discussed in Section 2, these models were chosen because they rely on design patterns shared by many renowned models for scene text recognition, as well as for their frequent roles as baselines in recent LPR research [12,17,19,20].

We implemented each model using the original framework or well-known public repositories associated with it. Specifically, we used Darknet[1] for the YOLO-based models (CR-NET and Fast-OCR). The multi-task models, Holistic-CNN and Multi-task-LR, were trained and evaluated using Keras. As for the remaining models, which were originally proposed for general scene text recognition, we used a fork[2] of the open source repository of Clova AI Research (PyTorch).

Here we list the hyperparameters employed in each framework for training the recognition models. These hyperparameters were determined based on existing research [1,2,19] and were further validated through experiments on the validation set. In Darknet, the parameters include: Stochastic Gradient Descent (SGD) optimizer, 90K iterations, a batch size of 64, and a learning rate of $[10^{-3}, 10^{-4}, 10^{-5}]$ with decay steps at 30K and 60K iterations. In Keras, we employed the Adam optimizer with an initial learning rate of 10^{-3} (ReduceL-ROnPlateau's patience of 5 and factor of 10^{-1}), a batch size of 64, and a patience value of 11 (patience indicates the number of epochs without improvement before training is stopped). In PyTorch, we used the following parameters: Adadelta optimizer with a decay rate of $\rho = 0.99$, 300K iterations, and a batch size of 128. The only modification we made to the models' architectures was adjusting the respective input layers to accommodate images with a width-to-height ratio of 3.

3.2   Datasets

Researchers have conducted experiments on various datasets to showcase the effectiveness of their systems in recognizing LPs from different regions [13,22, 26,36]. As shown in Table 2, we perform experiments using images from 12 public datasets commonly used to benchmark ALPR systems [18,19,24,38,46]. Each dataset was divided using standard splits, defined by the datasets' authors, or

---

[1] https://github.com/AlexeyAB/darknet.
[2] https://github.com/roatienza/deep-text-recognition-benchmark/.

following previous works [18, 22, 43] (when there is no standard split)³. Specifically, eight datasets were used both for training and evaluating the recognition models, while four were used exclusively for testing. The selected datasets exhibit substantial diversity in terms of image quantity, acquisition settings, image resolution, and LP layouts. As far as we know, no other work in ALPR research has conducted experiments using images from such a wide range of public datasets.

Table 2. The datasets employed in our experimental analysis, with '*' indicating those used exclusively for testing (i.e., in cross-dataset experiments). The "Chinese" layout denotes LPs assigned to vehicles registered in mainland China, while the "Taiwanese" layout corresponds to LPs issued for vehicles registered in the Taiwan region.

Dataset	Year	Images	LP Layout	Dataset	Year	Images	LP Layout
Caltech Cars [44]	1999	126	American	SSIG-SegPlate [9]	2016	2,000	Brazilian
EnglishLP [40]	2003	509	European	PKU* [45]	2017	2,253	Chinese
UCSD-Stills [5]	2005	291	American	UFPR-ALPR [21]	2018	4,500	Brazilian
ChineseLP [49]	2012	411	Chinese	CD-HARD* [36]	2018	102	Various
AOLP [15]	2013	2,049	Taiwanese	CLPD* [47]	2021	1,200	Chinese
OpenALPR-EU* [31]	2016	108	European	RodoSol-ALPR [19]	2022	20,000	Brazilian & Mercosur

The diversity of LP layouts across the selected datasets is depicted in Fig. 2, revealing considerable variations even among LPs from the same region. For instance, the EnglishLP and OpenALPR-EU datasets, both collected in Europe,

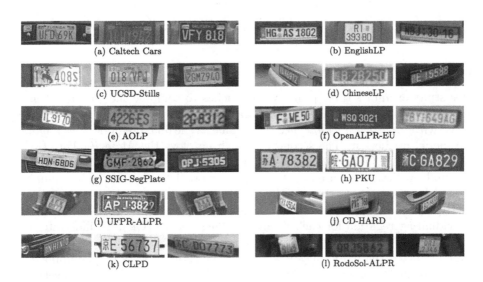

(a) Caltech Cars   (b) EnglishLP

(c) UCSD-Stills   (d) ChineseLP

(e) AOLP   (f) OpenALPR-EU

(g) SSIG-SegPlate   (h) PKU

(i) UFPR-ALPR   (j) CD-HARD

(k) CLPD   (l) RodoSol-ALPR

Fig. 2. Some LP images from the public datasets used in our experimental evaluation.

---

³ Detailed information on which images were used to train, validate and test the models can be accessed at https://raysonlaroca.github.io/supp/lpr-model-fusion/.

include images of LPs with notable distinctions in colors, aspect ratios, symbols (e.g., coats of arms), and the number of characters. Furthermore, certain datasets encompass LPs with two rows of characters, shadows, tilted orientations, and at relatively low spatial resolutions.

We explored various data augmentation techniques to ensure a balanced distribution of training images across different datasets. These techniques include random cropping, the introduction of random shadows, grayscale conversion, and random perturbations of hue, saturation, and brightness. Additionally, to counteract the propensity of recognition models to memorize sequence patterns encountered during training [8,10,46], we generated many synthetic LP images by shuffling the character positions on each LP (using the labels provided in [22]). Examples of these generated images are shown in Fig. 3.

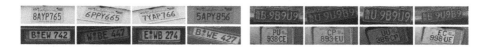

Fig. 3. Examples of LP images we created to mitigate overfitting. Within each group, the image on the left is the original, while the remaining ones are artificially generated counterparts. Various transformations were applied to enhance image variability.

To further mitigate the inherent biases present in public datasets [24], we expanded our training set by including 772 images from the internet. These images were annotated and made available by Ref. [22]. This supplementary dataset comprises 257 American LPs, 347 Chinese LPs, and 178 European LPs.

3.3   Fusion Approaches

This study examines three primary approaches to combine the outputs of multiple recognition models. The first approach involves selecting the sequence predicted with the *Highest Confidence* (HC) value as the final prediction, even if only one model predicts it. The second approach employs the *Majority Vote* (MV) rule to aggregate the sequences predicted by the different models. In other words, the final prediction is based on the sequence predicted by the largest number of models, disregarding the confidence values associated with each prediction. Lastly, the third approach follows a similar *Majority Vote* rule but performs individual aggregation for each *Character Position* (MVCP). To illustrate, the characters predicted in the first position are analyzed separately, and the character predicted the most times is selected. The same process is then applied to each subsequent character position until the last one. Ultimately, the selected characters are concatenated to form the final string.

One concern that arises when employing majority vote-based strategies is the potential occurrence of a tie. Let's consider a scenario where an LP image is processed by five recognition models. Two models predict "ABC-123," two models predict "ABC-124," and the remaining model predicts "ABC-125." In this case, a tie occurs between "ABC-123" and "ABC-124." To address this, we assess two tie-breaking approaches for each majority vote strategy: (i) selecting the prediction made with higher confidence among the tied predictions as the correct one, and (ii) selecting the prediction made by the "best model" as the correct one. In this study, for simplicity, we consider the best model the one that performs best individually across all datasets. However, in a more practical scenario, the chosen model could be the one known to perform best in the specific implementation scenario (e.g., one model may be the most robust for recognizing tilted LPs while another model may excel at handling low-resolution or noisy images). We intuitively use the acronym MV–HC to refer to the majority vote approach in which ties are broken by selecting the prediction made with the highest confidence value. Similarly, MV–BM refers to the majority vote approach in which ties are resolved by choosing the prediction made by the best model. The MVCP approaches follow a similar naming convention (MVCP–HC and MVCP–BM).

It is important to mention that when conducting fusion based on the highest confidence, we consider the confidence values derived directly from the models' outputs, even though some of them tend to make overconfident predictions. We carried out several experiments in which we normalized the confidence values of different models before fusing them, using various strategies such as weighted normalization based on the average confidence of each classifier's predictions. Somewhat surprisingly, these attempts did not yield improved results.

3.4 Performance Evaluation

In line with the standard practice in the literature, we report the performance of each experiment by calculating the ratio of correctly recognized LPs to the total number of LPs in the test set. An LP is considered correctly recognized only if all the characters on the LP are accurately identified, as even a single incorrectly recognized character can lead to the misidentification of the vehicle.

It is important to note that, although this work focuses on the LPR stage, the LP images used as input for the recognition models were not directly cropped from the ground truth. Instead, the YOLOv4 model [3] was employed to detect the LPs. This approach allows for a more accurate simulation of real-world scenarios, considering the imperfect nature of LP detection and the reduced robustness of certain recognition models when faced with imprecisely detected LP regions [10, 26]. As in [19], the results obtained using YOLOv4 were highly satisfactory. Considering detections with an Intersection over Union (IoU) ≥ 0.7 as correct, YOLOv4 achieved an average recall rate exceeding 99.7% in the test sets of the datasets used for training and validation, and 97.8% in the cross-dataset experiments. In both cases, the precision rates obtained were greater than 97%.

4   Results

Table 3 shows the recognition rates obtained on the disjoint test sets of the eight datasets used for training and validating the models. It presents the results

Table 3. Comparison of the recognition rates achieved across eight popular datasets by 12 models individually and through five different fusion strategies. Each model (rows) was trained once on the combined set of training images from all datasets and evaluated on the respective test sets (columns). The models are listed alphabetically, and the best recognition rates achieved in each dataset are shown in bold.

Approach	Caltech Cars # 46	EnglishLP # 102	UCSD-Stills # 60	ChineseLP # 161	AOLP # 687	SSIG-SegPlate # 804	UFPR-ALPR # 1,800	RodoSol-ALPR # 8,000	Average
CR-NET [37]	**97.8%**	94.1%	100.0%	97.5%	98.1%	97.5%	82.6%	†89.0%†	90.8%
CRNN [34]	93.5%	88.2%	91.7%	90.7%	97.1%	92.9%	68.9%	73.6%	87.1%
Fast-OCR [23]	93.5%	97.1%	100.0%	97.5%	98.1%	97.1%	81.6%	†56.7%†	90.2%
GRCNN [42]	93.5%	92.2%	93.3%	91.9%	97.1%	93.4%	66.6%	77.6%	88.2%
Holistic-CNN [39]	87.0%	75.5%	88.3%	95.0%	97.7%	95.6%	81.2%	94.7%	89.4%
Multi-task-LR [11]	89.1%	73.5%	85.0%	92.5%	94.9%	93.3%	72.3%	86.6%	85.9%
R²AM [25]	89.1%	83.3%	86.7%	91.9%	96.5%	92.0%	75.9%	83.4%	87.4%
RARE [35]	95.7%	94.1%	95.0%	94.4%	97.7%	94.0%	78.7%	78.7%	90.7%
Rosetta [4]	89.1%	82.4%	93.3%	93.8%	97.5%	94.4%	75.5%	89.0%	89.4%
STAR-Net [28]	95.7%	96.1%	95.0%	95.7%	97.8%	96.1%	78.8%	82.3%	92.2%
TRBA [2]	93.5%	91.2%	91.7%	93.8%	97.2%	97.3%	83.4%	80.6%	91.1%
ViTSTR-Base [1]	87.0%	88.2%	86.7%	96.9%	**99.4%**	95.6%	**89.7%**	95.6%	92.4%
Fusion HC (top 6)	**97.8%**	95.1%	96.7%	**98.1%**	99.0%	96.6%	90.9%	93.5%	96.0%
Fusion MV–BM (top 8)	**97.8%**	97.1%	100.0%	**98.1%**	**99.7%**	98.4%	92.7%	96.4%	97.5%
Fusion MV–BC (top 8)	**97.8%**	97.1%	100.0%	**98.1%**	**99.7%**	99.1%	92.3%	**96.5%**	**97.6%**
Fusion MVCP–BM (top 9)	95.7%	96.1%	100.0%	**98.1%**	99.6%	99.0%	**92.8%**	96.4%	97.2%
Fusion MVCP–BC (top 9)	**97.8%**	96.1%	100.0%	**98.1%**	99.6%	**99.3%**	92.5%	96.3%	97.5%

†Images from the RodoSol-ALPR dataset were not used for training the CR-NET and Fast-OCR models, as each character's bounding box needs to be labeled for training them.

Table 4. Average results obtained across the datasets by combining the output of the top N recognition models, ranked by accuracy, using five distinct strategies.

Models	HC	MV–BM	MV–HC	MVCP–BM	MVCP–HC
Top 1 (ViTSTR-Base)	92.4%	92.4%	92.4%	92.4%	92.4%
Top 2 (+ STAR-Net)	94.1%	92.4%	94.1%	92.4%	94.1%
Top 3 (+ TRBA)	94.2%	94.6%	94.9%	94.2%	94.2%
Top 4 (+ CR-NET)	95.2%	95.9%	96.3%	94.8%	95.9%
Top 5 (+ RARE)	95.5%	96.1%	96.6%	96.1%	96.2%
Top 6 (+ Fast-OCR)	**96.0%**	97.1%	97.0%	96.7%	96.9%
Top 7 (+ Rosetta)	95.4%	97.3%	97.2%	97.1%	97.0%
Top 8 (+ Holistic-CNN)	95.7%	**97.5%**	**97.6%**	96.1%	97.2%
Top 9 (+ GRCNN)	95.7%	95.5%	97.5%	**97.2%**	**97.5%**
Top 10 (+ R²AM)	95.5%	97.4%	97.2%	96.1%	96.6%
Top 11 (+ CRNN)	95.2%	97.1%	97.0%	96.5%	96.5%
Top 12 (+ Multi-task-LR)	95.0%	97.0%	97.0%	95.5%	96.5%

reached by each model individually, as well as the outcomes achieved through the fusion strategies outlined in Section 3.3. To improve clarity, Table 3 only includes the best results attained through model fusion. For a detailed breakdown of the results achieved by combining the outputs from the top 2 to the top 12 recognition models, refer to Table 4. The ranking of the models was determined based on their mean performance across the datasets (the ranking on the validation set was essentially the same, with only two models swapping positions).

Upon analyzing the results presented in Table 3, it becomes evident that model fusion has yielded substantial improvements. Specifically, the highest average recognition rate increased from 92.4% (ViTSTR-Base) to 97.6% by combining the outputs of multiple recognition models (MV–HC). While each model individually obtained recognition rates below 90% for at least one dataset (three on average), all fusion strategies surpassed the 90% threshold across all datasets. Remarkably, in most cases, fusion led to recognition rates exceeding 95%.

The significance of conducting experiments on multiple datasets becomes apparent as we observe that the best overall model (ViTSTR-Base) did not achieve the top result in five of the eight datasets. Notably, it exhibited relatively poor performance on the Caltech Cars, EnglishLP, and UCSD-Stills datasets. We attribute this to two primary reasons: (i) these datasets are older, containing fewer training images, which seems to impact certain models more than others (as explained in Section 3.2, we exploited data augmentation techniques to mitigate this issue); and (ii) these datasets were collected in the United States and Europe, regions known for having a higher degree of variability in LP layouts compared to the regions where the other datasets were collected (specifically, Brazil, mainland China, and Taiwan). It is worth noting that we included these datasets in our experimental setup, despite their limited number of images, precisely because they provide an opportunity to uncover such valuable insights.

Basically, by analyzing the results reported for each dataset individually, we observe that combining the outputs of multiple models does not necessarily lead to significantly improved performance compared to the best model in the ensemble. Instead, it reduces the likelihood of obtaining poor performance. This phenomenon arises because diverse models tend to make different errors for each

Fig. 4. Predictions obtained in eight LP images by multiple models individually and through the best fusion approach. Although we only show the predictions from the top 5 models for better viewing, it is noteworthy that in these particular cases, fusing the top 8 models (the optimal configuration) yielded identical predictions. The confidence for each prediction is indicated in parentheses, and any errors are highlighted in red. (Color figure online)

sample, but generally concur on correct classifications [32]. Illustrated in Fig. 4 are representative examples of predictions made by multiple models and the MV–HC fusion strategy on various LP images. It is remarkable that model fusion can produce accurate predictions even in cases where most models exhibit prediction errors. To clarify, with the MV–HC approach, this occurs when each incorrect sequence is predicted fewer times than the correct one, or in the case of a tie, the correct sequence is predicted with higher confidence.

Shifting our attention back to Table 4, we note that the majority vote-based strategies yielded comparable results, with the sequence-level approach (MV) performing marginally better for a given number of combined models. Our analysis indicates that this difference arises in cases where a model predicts one character more or one character less, impacting the majority vote by character position (MVCP) approach relatively more. Conversely, selecting the prediction with the highest confidence (HC) consistently led to inferior results. This can be attributed to the general tendency of all models to make incorrect predictions also with high confidence, as demonstrated in Fig. 4.

A growing number of authors [19,24,43,46] have stressed the importance of also evaluating LPR models in a cross-dataset fashion, as it more accurately simulates real-world scenarios where new cameras are regularly being deployed in new locations without existing systems being retrained as often. Taking this into account, we present in Table 5 the results obtained on four distinct datasets, none of which were used during the training of the models[4]. These particular datasets are commonly employed for this purpose [6,18,22,38,50].

Table 5. Comparison of the results achieved in cross-dataset setups by 12 models individually and through five different fusion strategies. The models are listed alphabetically, with the highest recognition rates attained for each dataset highlighted in bold.

Approach	Dataset # LPs				Average
	OpenALPR-EU # 108	PKU # 2,253	CD-HARD # 104	CLPD # 1,200	
CR-NET [37]	96.3%	99.1%	58.7%	94.2%	87.1%
CRNN [34]	93.5%	98.2%	31.7%	89.0%	78.1%
Fast-OCR [23]	**97.2%**	**99.2%**	**59.6%**	94.4%	**87.6%**
GRCNN [42]	87.0%	98.6%	38.5%	87.7%	77.9%
Holistic-CNN [39]	89.8%	98.6%	11.5%	90.2%	72.5%
Multi-task-LR [11]	85.2%	97.4%	10.6%	86.8%	70.0%
R²AM [25]	88.9%	97.1%	20.2%	88.2%	73.6%
RARE [35]	94.4%	98.3%	37.5%	92.4%	80.7%
Rosetta [4]	90.7%	97.2%	14.4%	86.9%	72.3%
STAR-Net [28]	**97.2%**	99.1%	48.1%	93.3%	84.4%
TRBA [2]	93.5%	98.5%	35.6%	90.9%	79.6%
ViTSTR-Base [1]	89.8%	98.4%	22.1%	93.1%	75.9%
Fusion HC (*top 6*)	95.4%	99.2%	48.1%	94.9%	84.4%
Fusion MV–BM (*top 8*)	**99.1%**	**99.7%**	**65.4%**	**97.0%**	**90.3%**
Fusion MV–HC (*top 8*)	**99.1%**	**99.7%**	**65.4%**	96.3%	90.1%
Fusion MVCP–BM (*top 9*)	95.4%	**99.7%**	54.8%	95.5%	86.3%
Fusion MVCP–HC (*top 9*)	97.2%	**99.7%**	57.7%	95.9%	87.6%

---

[4] To train the models, we excluded the few images from the ChineseLP dataset that are also found in CLPD (both datasets include internet-sourced images [20]).

These experiments provide further support for the findings presented earlier in this section. Specifically, both strategies that rely on a majority vote at the sequence level (MV–BM and MV–HC) outperformed the others significantly; the most notable performance gap was observed in the CD-HARD dataset, known for its challenges due to the predominance of heavily tilted LPs and the wide variety of LP layouts (as shown in Fig. 2). Interestingly, in this cross-dataset scenario, the MV–BM strategy exhibited slightly superior performance compared to MV–HC. Surprisingly, the HC approach failed to yield any improvements in results on any dataset, indicating that the models made errors with high confidence even on LP images extracted from datasets that were not part of their training.

While our primary focus lies on investigating the improvements in recognition rates achieved through model fusion, it is also pertinent to examine its impact on runtime. Naturally, certain applications might favor combining fewer models to attain a moderate improvement in recognition while minimizing the increase in the system's running time. With this in mind, Table 6 presents the number of frames per second (FPS) processed by each model independently and when incorporated into the ensemble. In addition to combining the models based on their average recognition rate across the datasets, as done in the rest of this section, we also explore combining them based on their processing speed.

Table 6. The number of FPS processed by each model independently and when incorporated into the ensembles. On the left, the models are ranked based on their results across the datasets, while on the right they are ranked according to their speed. The reported time, measured in milliseconds per image, represents the average of 5 runs.

Models (ranked by accuracy)	MV-HC	Individual		Fusion		Models (ranked by speed)	MV-HC	Individual		Fusion	
		Time	FPS	Time	FPS			Time	FPS	Time	FPS
Top 1 (ViTSTR-Base)	92.4%	7.3	137	7.3	137	Top 1 (Multi-task-LR)	85.9%	2.3	427	2.3	427
Top 2 (+ STAR-Net)	94.1%	7.1	141	14.4	70	Top 2 (+ Holistic-CNN)	90.2%	2.5	399	4.9	206
Top 3 (+ TRBA)	94.9%	16.9	59	31.3	32	Top 3 (+ CRNN)	91.1%	2.9	343	7.8	129
Top 4 (+ CR-NET)	96.3%	5.3	189	36.6	27	Top 4 (+ Fast-OCR)	95.4%	3.0	330	10.8	93
Top 5 (+ RARE)	96.6%	13.0	77	49.6	20	Top 5 (+ Rosetta)	96.0%	4.6	219	15.4	65
Top 6 (+ Fast-OCR)	97.0%	3.0	330	52.6	19	Top 6 (+ CR-NET)	96.6%	5.3	189	20.7	48
Top 7 (+ Rosetta)	97.2%	4.6	219	57.2	18	Top 7 (+ STAR-Net)	96.9%	7.1	141	27.8	36
Top 8 (+ Holistic-CNN)	97.6%	2.5	399	59.7	17	Top 8 (+ ViTSTR-Base)	96.9%	7.3	137	35.0	29
Top 9 (+ GRCNN)	97.5%	8.5	117	68.2	15	Top 9 (+ GRCNN)	97.1%	8.5	117	43.6	23
Top 10 (+ R^2AM)	97.2%	15.9	63	84.2	12	Top 10 (+ RARE)	97.1%	13.0	77	56.6	18
Top 11 (+ CRNN)	97.0%	2.9	343	87.1	11	Top 11 (+ R^2AM)	97.1%	15.9	63	72.5	14
Top 12 (+ Multi-task-LR)	97.0%	2.3	427	89.4	11	Top 12 (+ TRBA)	97.1%	16.9	59	89.4	11

Remarkably, fusing the outputs of the three fastest models results in a lower recognition rate (91.1%) than using the best model alone (92.4%). Nevertheless, as more methods are included in the ensemble, the gap reduces considerably. From this observation, we can infer that if attaining the utmost recognition rate across various scenarios is not imperative, it becomes more advantageous to combine fewer but faster models, as long as they perform satisfactorily individually. According to Table 6, combining 4–6 fast models appears to be the optimal choice for striking a better balance between speed and accuracy.

5   Conclusions

This paper examines the potential improvements in LPR results by fusing the outputs from multiple recognition models. Distinguishing itself from prior studies, our research explores a wide range of models and datasets in the experiments. We combined the outputs of different models through straightforward approaches such as selecting the most confident prediction or through majority vote (both at sequence and character levels), demonstrating the substantial benefits of fusion approaches in both intra- and cross-dataset experimental setups.

In the traditional intra-dataset setup, where we explored eight datasets, the mean recognition rate experienced a significant boost, rising from 92.4% achieved by the best model individually to 97.6% when leveraging model fusion. Essentially, we demonstrate that fusing multiple models reduces considerably the likelihood of obtaining subpar performance on a particular dataset. In the more challenging cross-dataset setup, where we explored four datasets, the mean recognition rate increased from 87.6% to rates surpassing 90%. Notably, the optimal fusion approach in both setups was via majority vote at sequence level.

We also conducted an evaluation to analyze the speed/accuracy trade-off in the final approach by varying the number of models included in the ensemble. For this assessment, we ranked the models in two distinct ways: one based on their recognition results and the other based on their efficiency. The findings led us to conclude that for applications where the recognition task can tolerate some additional time, though not excessively, an effective strategy is to combine 4–6 fast models. Employing this approach significantly enhances the recognition results while maintaining the system's efficiency at an acceptable level.

Acknowledgments. We thank the support of NVIDIA Corporation with the donation of the Quadro RTX 8000 GPU used for this research.

References

1. Atienza, R.: Vision transformer for fast and efficient scene text recognition. In: International Conference on Document Analysis and Recognition, pp. 319–334 (2021)
2. Baek, J., et al.: What is wrong with scene text recognition model comparisons? Dataset and model analysis. In: IEEE/CVF International Conference on Computer Vision (ICCV), pp. 4714–4722 (2019)
3. Bochkovskiy, A., Wang, C.Y., Liao, H.Y.M.: YOLOv4: optimal speed and accuracy of object detection, pp. 1–14. arXiv preprint arXiv:2004.10934 (2020)
4. Borisyuk, F., Gordo, A., Sivakumar, V.: Rosetta: large scale system for text detection and recognition in images. In: ACM SIGKDD International Conference on Knowledge Discovery & Data Mining, pp. 71–79 (2018)
5. Dlagnekov, L.: UCSD/Calit2 car license plate, make and model database (2005). https://www.belongielab.org/car_data.html
6. Fan, X., Zhao, W.: Improving robustness of license plates automatic recognition in natural scenes. IEEE Trans. Intell. Transp. Syst. **23**(10), 18845–18854 (2022)

7. Gao, Y., et al.: GroupPlate: toward multi-category license plate recognition. IEEE Trans. Intell. Transp. Syst. **24**(5), 5586–5599 (2023)
8. Garcia-Bordils, S., et al.: Out-of-vocabulary challenge report. In: European Conference on Computer Vision, TiE: Text in Everything Workshop, pp. 1–17 (2022)
9. Gonçalves, G.R., Silva, S.P.G., Menotti, D., Schwartz, W.R.: Benchmark for license plate character segmentation. J. Electron. Imaging **25**(5), 053034 (2016)
10. Gonçalves, G.R., et al.: Real-time automatic license plate recognition through deep multi-task networks. In: Conference on Graphics, Patterns and Images (SIBGRAPI), pp. 110–117 (Oct 2018)
11. Gonçalves, G.R., et al.: Multi-task learning for low-resolution license plate recognition. In: Iberoamerican Congress on Pattern Recognition, pp. 251–261 (Oct 2019)
12. Gong, Y., et al.: Unified Chinese license plate detection and recognition with high efficiency. J. Vis. Commun. Image Represent. **86**, 103541 (2022)
13. Henry, C., Ahn, S.Y., Lee, S.: Multinational license plate recognition using generalized character sequence detection. IEEE Access **8**, 35185–35199 (2020)
14. Hsu, G.S., Ambikapathi, A., Chung, S.L., Su, C.P.: Robust license plate detection in the wild. In: IEEE International Conference on Advanced Video and Signal Based Surveillance (AVSS), pp. 1–6 (2017)
15. Hsu, G.S., Chen, J.C., Chung, Y.Z.: Application-oriented license plate recognition. IEEE Trans. Veh. Technol. **62**(2), 552–561 (2013)
16. Izidio, D.M.F., et al.: An embedded automatic license plate recognition system using deep learning. Des. Autom. Embed. Syst. **24**(1), 23–43 (2020)
17. Kabiraj, A., Pal, D., Ganguly, D., Chatterjee, K., Roy, S.: Number plate recognition from enhanced super-resolution using generative adversarial network. Multimedia Tools Appli. **82**(9), 13837–13853 (2023)
18. Ke, X., Zeng, G., Guo, W.: An ultra-fast automatic license plate recognition approach for unconstrained scenarios. IEEE Trans. Intell. Transp. Syst. **24**(5), 5172–5185 (2023)
19. Laroca, R., Cardoso, E.V., Lucio, D.R., Estevam, V., Menotti, D.: On the cross-dataset generalization in license plate recognition. In: International Conference on Computer Vision Theory and Applications (VISAPP), pp. 166–178 (2022)
20. Laroca, R., Estevam, V., Britto Jr., A.S., Minetto, R., Menotti, D.: Do we train on test data? the impact of near-duplicates on license plate recognition. In: International Joint Conference on Neural Networks (IJCNN), pp. 1–8 (2023)
21. Laroca, R., Severo, E., Zanlorensi, L.A., Oliveira, L.S., et al.: A robust real-time automatic license plate recognition based on the YOLO detector. In: International Joint Conference on Neural Networks (IJCNN), pp. 1–10 (2018)
22. Laroca, R., Zanlorensi, L., Gonçalves, G., Todt, E., Schwartz, W., Menotti, D.: An efficient and layout-independent automatic license plate recognition system based on the YOLO detector. IET Intel. Transport Syst. **15**(4), 483–503 (2021)
23. Laroca, R., et al.: Towards image-based automatic meter reading in unconstrained scenarios: a robust and efficient approach. IEEE Access **9**, 67569–67584 (2021)
24. Laroca, R., et al.: A first look at dataset bias in license plate recognition. In: Conference on Graphics, Patterns and Images (SIBGRAPI), pp. 234–239 (2022)
25. Lee, C., Osindero, S.: Recursive recurrent nets with attention modeling for OCR in the wild. In: IEEE/CVF Conference on Computer Vision and Pattern Recognition (CVPR), pp. 2231–2239 (2016)
26. Lee, Y., et al.: License plate detection via information maximization. IEEE Trans. Intell. Transp. Syst. **23**(9), 14908–14921 (2022)

27. Liu, Q., Chen, S.L., Li, Z.J., Yang, C., Chen, F., Yin, X.C.: Fast recognition for multidirectional and multi-type license plates with 2D spatial attention. In: International Conference on Document Analysis and Recognition (ICDAR), pp. 125–139 (2021)
28. Liu, W., Chen, C., Kwan-Yee K. Wong, Z.S., Han, J.: STAR-Net: a spatial attention residue network for scene text recognition. In: British Machine Vision Conference (BMVC), pp. 1–13 (Sept 2016)
29. Mokayed, H., Shivakumara, P., Woon, H.H., Kankanhalli, M., Lu, T., Pal, U.: A new DCT-PCM method for license plate number detection in drone images. Pattern Recogn. Lett. **148**, 45–53 (2021)
30. Nascimento, V., et al.: Super-resolution of license plate images using attention modules and sub-pixel convolution layers. Comput. Graph. **113**, 69–76 (2023)
31. OpenALPR: OpenALPR-EU dataset (2016). https://github.com/openalpr/benchmarks/tree/master/endtoend/eu
32. Polikar, R.: Ensemble learning. Ensemble Machine Learning: Methods and Applications, pp. 1–34. Springer, New York (2012). https://doi.org/10.1007/978-1-4419-9326-7
33. Schirrmacher, F., Lorch, B., Maier, A., Riess, C.: Benchmarking probabilistic deep learning methods for license plate recognition. IEEE Trans. Intell. Transp. Syst. **24**(9), 9203–9216 (2023)
34. Shi, B., Bai, X., Yao, C.: An end-to-end trainable neural network for image-based sequence recognition and its application to scene text recognition. IEEE Trans. Pattern Anal. Mach. Intell. **39**(11), 2298–2304 (2017)
35. Shi, B., Wang, X., Lyu, P., Yao, C., Bai, X.: Robust scene text recognition with automatic rectification. In: IEEE/CVF Conference on Computer Vision and Pattern Recognition (CVPR), pp. 4168–4176 (2016)
36. Silva, S.M., Jung, C.R.: License plate detection and recognition in unconstrained scenarios. In: Ferrari, V., Hebert, M., Sminchisescu, C., Weiss, Y. (eds.) ECCV 2018. LNCS, vol. 11216, pp. 593–609. Springer, Cham (2018). https://doi.org/10.1007/978-3-030-01258-8_36
37. Silva, S.M., Jung, C.R.: Real-time license plate detection and recognition using deep convolutional neural networks. J. Vis. Commun. Image Represent. 102773 (2020)
38. Silva, S.M., Jung, C.R.: A flexible approach for automatic license plate recognition in unconstrained scenarios. IEEE Trans. Intell. Transp. Syst. **23**(6), 5693–5703 (2022)
39. Špaňhel, J., et al.: Holistic recognition of low quality license plates by CNN using track annotated data. In: IEEE International Conference on Advanced Video and Signal Based Surveillance (AVSS), pp. 1–6 (2017)
40. Srebrić, V.: EnglishLP database (2003). https://www.zemris.fer.hr/projects/LicensePlates/english/baza_slika.zip
41. Terven, J., Cordova-Esparza, D.: A comprehensive review of YOLO: From YOLOv1 and beyond, pp. 1–33. arXiv preprint arXiv:2304.00501 (2023)
42. Wang, J., Hu, X.: Gated recurrent convolution neural network for OCR. In: International Conference on Neural Information Processing Systems (NeurIPS), pp. 334–343 (2017)
43. Wang, Y., Bian, Z.P., Zhou, Y., Chau, L.P.: Rethinking and designing a high-performing automatic license plate recognition approach. IEEE Trans. Intell. Transp. Syst. **23**(7), 8868–8880 (2022)
44. Weber, M.: Caltech Cars (1999). https://data.caltech.edu/records/20084

45. Yuan, Y., Zou, W., Zhao, Y., Wang, X., Hu, X., Komodakis, N.: A robust and efficient approach to license plate detection. IEEE Trans. Image Process. **26**(3), 1102–1114 (2017)
46. Zeni, L.F., Jung, C.R.: Weakly supervised character detection for license plate recognition. In: Conference on Graphics, Patterns and Images, pp. 218–225 (2020)
47. Zhang, L., Wang, P., Li, H., Li, Z., Shen, C., Zhang, Y.: A robust attentional framework for license plate recognition in the wild. IEEE Trans. Intell. Transp. Syst. **22**(11), 6967–6976 (2021)
48. Zhang, M., Liu, W., Ma, H.: Joint license plate super-resolution and recognition in one multi-task GAN framework. In: IEEE International Conference on Acoustics, Speech and Signal Processing (ICASSP), pp. 1443–1447 (2018)
49. Zhou, W., et al.: Principal visual word discovery for automatic license plate detection. IEEE Trans. Image Process. **21**(9), 4269–4279 (2012)
50. Zou, Y., et al.: A robust license plate recognition model based on Bi-LSTM. IEEE Access **8**, 211630–211641 (2020)

Enhancing Object Detection in Maritime Environments Using Metadata

Diogo Samuel Fernandes[1]($\boxtimes$), João Bispo[1], Luís Conde Bento[2,4],
and Mónica Figueiredo[2,3]

[1] Faculdade de Engenharia da Universidade do Porto, Porto, Portugal
up201806250@edu.fe.up.pt
[2] Politécnico de Leiria, Leiria, Portugal
[3] Instituto de Telecomunicações, Aveiro, Portugal
[4] Instituto de Sistemas e Robótica, Coimbra, Portugal

Abstract. Over the years, many solutions have been suggested in order to improve object detection in maritime environments. However, none of these approaches uses flight information, such as altitude, camera angle, time of the day, and atmospheric conditions, to improve detection accuracy and network robustness, even though this information is often available and captured by the UAV. This work aims to develop a network unaffected by image-capturing conditions, such as altitude and angle. To achieve this, metadata was integrated into the neural network, and an adversarial learning training approach was employed. This was built on top of the YOLOv7, which is a state-of-the-art realtime object detector. To evaluate the effectiveness of this methodology, comprehensive experiments and analyses were conducted. Findings reveal that the improvements achieved by this approach are minimal when trying to create networks that generalize more across these specific domains. The YOLOv7 mosaic augmentation was identified as one potential responsible for this minimal impact because it also enhances the model's ability to become invariant to these image-capturing conditions. Another potential cause is the fact that the domains considered (altitude and angle) are not orthogonal with respect to their impact on captured images. Further experiments should be conducted using datasets that offer more diverse metadata, such as adverse weather and sea conditions, which may be more representative of real maritime surveillance conditions. The source code of this work is publicly available at https://github.com/ipleiria-robotics/maritime-metadata-adaptation.

Keywords: Computer Vision · Remote Sensing · Maritime Surveillance · Domain Adaptation · Metadata

1 Introduction

UAVs' aerial images differ from those usually used on object detection tasks, which mostly include large objects and a horizontal angle of view. Drone images have different shooting angles, the objects to be detected are often very small, and the features to be extracted are sometimes ambiguous [1]. Images taken over the ocean suffer from

V. Vasconcelos et al. (Eds.): CIARP 2023, LNCS 14470, pp. 76–89, 2024.
https://doi.org/10.1007/978-3-031-49249-5_6

other effects, such as extreme glare, boat wakes, and white caps caused by waves [2]. Moreover, adverse weather conditions such as rain and fog will also degrade the quality of images, making it more challenging to perform object detection in this context. Alongside images, UAVs have other information available, such as altitude, camera angle, date, time, and location. This data is commonly referred to as metadata and can offer valuable details. The embedded devices on UAVs also have limited computational power, making it difficult to run some of the larger and more computationally intensive algorithms commonly used for object detection [3]. Making these algorithms achieve real-time inference in edge devices while maintaining an acceptable performance is essential in this application scenario.

This work aims to explore the detection of ships and other objects in the sea using images captured by UAVs and additional flight-related information that may be useful to improve the detection performance while keeping the complexity under control.

2 Related Work

2.1 Maritime Surveillance Datasets

UAV images pose challenges in image-based surveillance since they have a variable angle perspective and altitude, significantly impacting the size of objects to be detected. In 2019, Ribeiro et al. [2] presented a dataset containing images from UAVs captured on the coast of Lisbon. The images have different altitudes and shooting angles. The images also have glare, boat wakes, and waves, making detecting ships in this environment more challenging. Chen et al. [4] also created a dataset with images taken under different weather conditions, angles, and altitudes. It includes images captured in a foggy environment with many small ships, which can be challenging to detect. However, neither of these datasets includes flight-related information, such as altitude and shooting angles. SeaDronesSee [5] is a recent dataset published in 2022 that contains this information. The authors collected data and annotated over 54,000 frames from high altitudes and viewing angles from the videos. They evaluated state-of-the-art algorithms on their dataset and provided a baseline. The difficulty of gathering real data associated with the cost and legal limitations of using UAVs led to the creation of synthetic datasets, as pointed out by Kiefer et al. in their study [6].

2.2 Data-Level Techniques to Enhance Object Detection

Improving object detection requires changes to deep learning models and the training data. By preprocessing and restoring images, it is possible to minimize the effects of glare, fog, and other factors, enhancing the image quality. SRCYOLO [7] is an object detection network that tackles the issue of foggy images by incorporating Single Scale Retinex [8] defogging algorithm at the input level. However, Nie et al. [9] stated that image restoration techniques could degrade the images, leading to a worse detection of objects. To test this hypothesis, they synthesized hazy and low-light images. Then, they trained a neural network using three approaches: (1) without synthesized images and image preprocessing, (2) with image preprocessing only, and (3) using only synthesized images. The third approach performed better, while the other two produced nearly identical results.

2.3   Network-Level Techniques to Enhance Object Detection

One of the main challenges is that vessels and ships are usually captured from many different perspectives, making them appear smaller or larger due to their distance from the camera. If the object detection system were trained on a single scale of these images, the model might struggle to detect the vessel in the other perspectives accurately. Training on multiple scales allows the model to become more robust. HSF-Net [10] is a network that embeds feature maps in different scales to the same space using an HSF layer. The network is similar to the faster R-CNN, having a feature extraction network with multiple convolutional and pooling layers, an RPN, and a DN. The HSF layer is applied both to RPN and DN. This layer focuses on extracting the shape instead of detailed local features. Three scales are utilized to embed objects of varying sizes into the same dimensional feature space. Chen et al. [11] proposed and tested a similar method based on an optimized FPN, incorporated in a traditional RPN, and then mapped into a new feature space. The k-means algorithm based on Shape Similar Distance clustering is then used to obtain the initial anchor boxes. The proposed network outperformed previous models in complex scenarios, showing that it is appropriate for multi-scale and multi-target recognition and detecting small ships. These approaches can also be employed on one-stage detectors, resulting in better performance and faster processing. Zhang et al. [12] created a multi-layer convolution feature fusion (CFF-SDN). This model fuses shallow with deep features, which are then used for classification and regression.

In addition to multi-scale features, properly adjusted anchor boxes are crucial to ensure no missing objects. To address this, Hong et al. [13] improved the anchor boxes by using a linear scaling based on the k-means++ instead of the k-means clustering algorithm. They also used a Gaussian model to output the uncertainty of each prediction bounding box, improving the detection accuracy, and four anchors are assigned in the detection layer to increase the robustness of this model for detecting objects in multiple scales.

Researchers are also using attention modules to improve their networks. Li et al. [14] proposed an improved YOLOv3 tiny network with the CBAM attention module at the end of the backbone. The authors also used convolution layers instead of max-pooling and expanded the number of input channels for prediction, improving the detection of small objects. Chen et al. [15] proposed a Dilated Attention Module (DAM) to extract feature representations of ship targets. This lightweight attention module has a larger receptive field that can capture a wider variety of surrounding information and a residual connection, which helps discern small ship targets in harsher environments. Li et al. [16] use the Coordinate Attention (CA) module, putting it on the last layer of the backbone of the YOLOv5. The CA embeds positional information into channel attention, allowing the network to focus on large regions without introducing a large amount of processing. Later, the same researchers proposed a backbone with both convolutions and transformers [17]. In the neck, the feature maps are fused based on GhostNet, and the head employs three detectors of different scales to calculate the position and size of the objects. Transformer networks are also starting to appear, even though they have more considerable complexity than those based on convolutions. Sun et al. [3] uses the Swin-Transformer, a hierarchical transformer, as the backbone network for Faster R-CNN.

2.4 Metadata Integration

Some datasets contain images with different features and characteristics that make object detection harder, such as weather conditions (i.e., fog, rain, and sunny weather). Each of these conditions can be treated as a distinct domain. Training a model in a specific domain and ensuring its effective generalization to other domains is commonly known as domain adaptation [18]. Ganim et al. [19] were the first to propose a method using adversarial learning to perform domain adaptation. The architecture consists of a feature extractor that processes the input image, followed by a label predictor to classify the image and a domain classifier to determine the image domain. The gradients of the domain classifier are passed through a gradient reversal layer to make the feature extractor learn domain-invariant features. On the other hand, the class labels undergo normal backpropagation so that the network can learn how to classify each class correctly. This approach is also helpful when we want to make the network invariant to these domains, making the networks more robust. Li et al. [20] enhanced object detection performance in autonomous driving under adverse weather conditions using a similar methodology with adversarial learning. Zhenyu Wu et al. [21] also employed this adversarial learning methodology to enhance the model's ability to generalize effectively across diverse domains in drone images, which is a similar case to the one addressed in this work. This approach ensured that the underlying network focused on extracting only relevant features by becoming invariant to three distinct domains, which were altitude, angle, and weather.

3 Data Analysis

Since our approach focuses on using metadata to improve detection, we chose SeaDronesSee [5], which is a large dataset with images captured from UAV designed to help develop systems for search and rescue operations. The relevant metadata variables available were the altitude and the camera angle at which the images were taken. The authors of the dataset have supplied the data with pre-established divisions into training, testing, and validation sets.

The altitude represents the vertical distance between the aircraft and the sea level at the moment of image capture. The altitude at which the pictures were taken influences the object size, scale, and level of detail, meaning that higher altitudes provide a more challenging time at detection. Figure 1 provides valuable insights into the distribution of altitudes at which the images were captured. We divided the altitude domain into three intervals, low, medium, and high, each one with the same number of images. Each interval corresponds to a distinct altitude range and will be used by our network to become more robust.

On the other hand, the angle variable refers to the orientation from which the images were captured. Figure 2 shows the distribution of angles at which the images were captured. Similarly to the altitude, we divided the angle variable into three groups based on the respective camera angles. The groups encompassed angles between 0 and 35°, 36 to 65°, and from 66 to 90°.

During our analysis of altitude and angle variables in the dataset, an interesting pattern came to light. It became evident that images captured at higher altitudes frequently

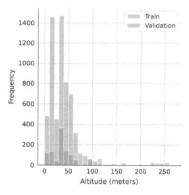
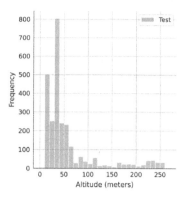

Fig. 1. Image altitude distribution.

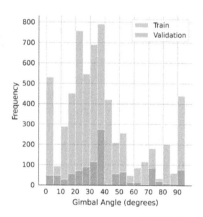
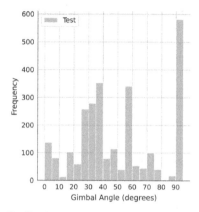

Fig. 2. Image angle distribution.

showed a downward angle of the camera, while images taken at lower altitudes tended to have a more horizontal camera orientation. This implies that the altitude and camera angle variables are non-orthogonal.

4 Proposed Architecture

In this study, our objective is to investigate the potential impact of variations, specifically the altitude and angle, on the performance of object detection algorithms in maritime environments. Our goal is to have a model that preserves object-related information while discarding or de-emphasizing the influence of these variations unrelated to the objects. This leads to a problem in which:

– A domain classification network receives latent representations from the feature extractor and will try to predict the domain associated with a given image.
– An object detection network receives latent representations from the feature extractor and will try to predict the objects associated with a given image.

– The feature extractor will attempt to deceive the domain classification network while providing high-quality representations to ensure optimal performance of the object detection network.

UAVs come with diverse hardware setups, many of which use the NVIDIA Jetson platform, capable of running YOLOv7 [22]. With this in mind, we used YOLOv7 as the foundation of our model, where the feature extractor consists of the first 40 layers of the backbone and the object detection is built with the remaining 10 layers of the backbone, along with the head. The domain classification consists of convolutional and fully connected layers. However, one problem with using YOLOv7 is the high number of image augmentations responsible for making this network more robust and high-performing, making it impossible to discern which domain the image belongs to. Because of this, augmented images are only used for object detection and images without augmentations for domain classification. In Fig. 3, it is possible to see the high-level diagram of our architecture.

Two techniques were used to assess the network's performance at inference time. The first method involved conducting end-to-end inference using the model trained on either a single domain or multiple domains using an adversarial approach. The second method of inference was using a model ensemble. Each base model was trained in one domain and then combined to produce a final prediction for unseen data. This approach was chosen to avoid potential gradient cancellation during backpropagation when using multiple-domain classification networks.

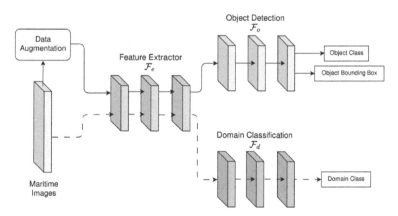

Fig. 3. Diagram of our architecture with image processing.

5 Training Methodology

Acknowledging that this deliberate emphasis on making the model invariant to domain characteristics may have potential drawbacks is essential. The model may inadvertently overlook being invariant to these variables and struggle to detect objects, which is the main task. To mitigate the potential impact of these inherent characteristics, we built a custom training loop that comprises three distinct stages, each contributing to the overall process of training our model.

5.1 First Stage

During the first stage of our methodology, we focus on training the domain classifier to distinguish between different domains. As depicted in Fig. 4, the object detection network is not used, and the feature extractor is frozen. We train the domain classification until we meet a minimum of 70% validation accuracy or 90% training accuracy or after a maximum number of epochs to avoid overfitting. The input data is processed through the feature extractor, and the resulting output is subsequently subjected to image domain classification, outputting the predicted probability p for each domain. The loss that the domain classification tries to minimize is shown in Eq. (1). Here, gt_i represents the true probability of class i, and p_i represents the predicted probability of class i. Following this process, we obtain a robust domain classifier network capable of accurately predicting the domain labels.

$$L_{domain}(gt, p) = - \sum_{i=1}^{N} gt_i \log(p_i) \tag{1}$$

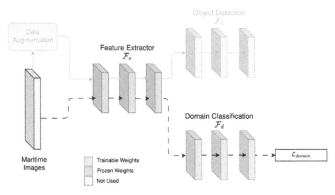

Fig. 4. First stage of our training loop.

5.2 Second Stage

During the second stage, the domain classifier network is frozen. The feature extractor is updated by the adversarial and object detection loss, while the object detection network is updated just by the object detection loss. The input data is processed again through the feature extractor, and the resulting output is passed through two subsequent branches. Augmented images are fed through the feature extractor and then to the object module, while non-augmented images are fed through the feature extractor and then to the domain classifier, similar to the first stage of the training. This can be seen in Fig. 5.

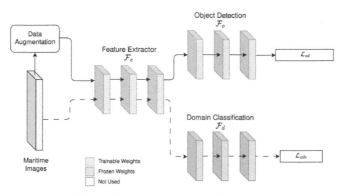

Fig. 5. Second stage of our training loop.

This helps the encoder to learn features in a way that the predictions of the object detection are as accurate as possible, and the predictions of the domain classification are kept close to a uniform distribution, meaning that this domain classifier is not better than random guessing the corresponding domain. For this, we use the KL divergence as the adversarial domain loss. The KL loss is depicted in Eq. (2), being P_i the probability of domain i according to the predicted distribution P, and Q_i represents the probability of domain i according to the uniform target distribution Q. By minimizing the KL divergence, we encourage the predicted probabilities to approach the uniform distribution.

$$\mathcal{L}_{adv}(P||Q) = \sum_{i}^{N} P_i \log\left(\frac{P_i}{Q_i}\right) \tag{2}$$

The loss regarding object detection is the one used in the YOLOv7, and we denote this loss as L_{od}, which is composed of the bounding box, objectness, and classification loss. The loss function is presented in Eq. (3), which is the sum of the object detection loss L_{od} and the adversarial loss L_{adv}, which is multiplied by a weighting factor λ. By jointly training these branches, the model will learn domain invariant features.

$$L = L_{od} + \lambda L_{adv} \tag{3}$$

5.3 Third Stage

During the third stage, the main focus is updating the model weights based solely on the object detection task, as depicted in Fig. 6. This means that the training process exclusively relies on the object detection loss without considering the adversarial domain loss or any domain-related factors. The reason for using only object detection in this stage is to ensure that the feature extractor learns to extract relevant features from the images and objects in the dataset. This stage continues for a specific number of epochs. After completing this stage, the process returns to the first stage, and the weights of the domain classifier, optimizer, and scheduler are reset. By resetting the domain classification module, we avoid potential bias or interference from previous domain-specific information.

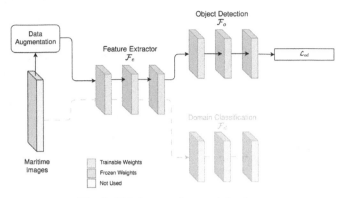

Fig. 6. Third stage of our training loop.

6   Results and Discussion

We start our work improving the baseline to ensure a solid enough validation for our proposed methodology. We evolved the YOLOv7 hyperparameters for 350 iterations. The image size used was 640, although bigger sizes may have yielded better results. Unfortunately, we used an NVIDIA A2000 for training, which restricted our ability to work with larger image sizes. To ensure that our results are reliable and not dependent on seed randomness, we run the models on different seeds. Table 1 shows the model performance using evolved hyperparameters.

Table 1. Baseline Performance.

Model	mAP	mAP@50	mAP@75	AR1	AR10
Seed 0	43.02	71.86	45.00	38.26	50.17
Seed 42	42.55	71.42	44.29	37.54	49.61
Seed 821	42.18	70.81	44.46	37.08	49.25
Seed 1765	42.53	71.25	44.94	37.52	49.41
Average	**42.57**	**71.34**	**44.67**	**37.6**	**49.61**
Standard Deviation	0.34	0.43	0.35	0.49	0.40

6.1   Domain Adversarial Learning

We started by making the model become invariant to the different altitude intervals. We performed some experiments using different combinations of N, the number of epochs that the model will train before using the domain adversarial network again, and λ, the weighting factor for our adversarial loss. We started by experimenting with $N = 1$, but the obtained results were unsatisfactory and subsequently disregarded, leading us to choose to test with 4 and 8, as can be seen in Table 2. We conducted additional runs with the best model using different seeds, as shown in Table 3, obtaining an average mAP of 43.05%, which signifies a marginal increase of 0.48. The lower gain at mAP@50 compared to mAP@75 can happen because objects with small dimensions can be easily missed.

Table 2. Performance using altitude domain.

N	λ	mAP	mAP@50	mAP@75	AR1	AR10
8	0.001	**43.85**	**74.28**	45.85	**38.39**	**51.57**
	0.005	42.88	71.75	45.30	37.45	49.67
	0.01	42.75	72.24	44.20	37.61	49.93
4	0.001	43.17	73.47	45.07 **46.16**	37.89	50.50
	0.005	43.09	71.60	44.48	37.77	50.23
	0.01	42.52	71.95		37.63	49.58

Table 3. Performance using the best parameters using altitude.

Model	mAP	mAP@50	mAP@75	AR1	AR10
Seed 0	43.85	74.28	45.85	38.39	51.57
Seed 42	42.17	71.74	43.55	37.26	49.65
Seed 821	42.96	72.26	44.75	37.48	49.88
Seed 1765	43.23	72.70	46.00	37.70	50.25
Average	43.05	72.75	45.04	37.71	50.34
Standard Deviation	0.70	1.10	1.14	0.49	0.86
Baseline Average	42.57	71.34	44.67	37.60	49.61
Improvement over Baseline Average	**0.48**	**1.41**	**0.37**	**0.11**	**0.73**

The same tests were employed for the angle variable and are depicted in Table 4 where the highest achieved performance was 43.77% when using $N = 8$ and $\lambda = 0.005$. The average mAP of the best model in different seeds was 43.24% which represents an increment of 0.67, as can be seen in Table 5. Similar to the altitude results, we observed higher gains at mAP@50, with an improvement of 1.30. The increase in performance at mAP@75 is relatively small, with an improvement of only 0.47.

Testing with both domains simultaneously is expected to improve performance. For this approach, two domain classification modules were used, one for the altitude and one for the angle, and a different λ was employed for each domain. The results using the best combinations can be found in Table 6 and show lower gains regarding using only one domain. One possible factor for this occurrence could be the mutual exclusion of gradients since the domains considered are not orthogonal, leading to the mutual exclusion of gradients.

The ensemble model employed in this study also tries to be invariant to multiple domains at the same time. We combine the weights from different pre-trained models using the adversarial learning technique. We integrated the weights obtained from the

Table 4. Performance using angle domain.

N	λ	mAP	mAP@50	mAP@75	AR1	AR10
8	0.001	42.96	71.83	45.54	37.76	50.29
	0.005	**43.77**	**74.15**	**45.56**	**38.84**	**51.23**
	0.01	43.02	72.05	45.45	37.63	50.07
4	0.001	43.19	72.72	45.07	37.92	50.09
	0.005	43.36	72.39	45.65	38.27	50.5
	0.01	40.91	69.37	41.82	36.10	48.34

Table 5. Performance using the best parameters using angle.

Model	mAP	mAP@50	mAP@75	AR1	AR10
Seed 0	43.77	74.15	45.56	38.84	51.23
Seed 42	43.51	73.45	45.34	37.91	50.45
Seed 821	42.41	71.13	44.08	37.62	50.01
Seed 1765	43.28	71.83	45.56	38.18	50.47
Average	43.24	72.64	45.14	38.14	50.54
Standard Deviation	0.59	1.40	0.71	0.52	0.51
Baseline Average	42.57	71.34	44.67	37.60	49.61
Improvement over Baseline Average	**0.67**	**1.30**	**0.47**	**0.54**	**0.93**

Table 6. Performance using the best parameters using multiple domains.

Model	mAP	mAP@50	mAP@75	AR1	AR10
Seed 0	43.61	73.18	45.58	38.15	50.64
Seed 42	42.83	72.24	44.52	37.64	50.07
Seed 821	42.57	71.13	44.77	36.87	49.38
Seed 1765	42.49	71.40	45.03	37.33	49.44
Average	42.88	71.99	44.98	37.50	49.89
Standard Deviation	0.51	0.93	0.46	0.54	0.60
Baseline Average	42.57	71.34	44.67	37.60	49.61
Improvement over Baseline Average	**0.31**	**0.65**	**0.31**	**−0.1**	**0.28**

models that were solely trained on a single domain, either on altitude or angle. To illustrate this with an example, we used the weights from the best angle model of seed 0 and the weights from the best altitude model of seed 0 to form an ensemble model. The results of this approach are presented in Table 7. This ensemble model yielded the best results regarding all techniques. By combining the strengths of multiple models, we enhanced overall performance and achieved moderate gains with a 1.44 improvement on mAP and an improvement of 3.26 on mAP@50 over the baseline model.

Table 7. Performance using ensemble model.

Model	mAP	mAP@50	mAP@75	AR1	AR10
Seed 0	45.12	76.57	46.78	39.93	53.47
Seed 42	43.62	73.87	45.22	38.54	51.44
Seed 821	43.41	73.53	44.76	38.16	51.06
Seed 1765	43.90	74.44	46.10	38.64	51.43
Average	44.01	74.60	45.72	38.82	51.85
Standard Deviation	0.77	1.36	0.90	0.77	1.09
Baseline Average	42.57	71.34	44.67	37.60	49.61
Improvement over Baseline Average	**1.44**	**3.26**	**1.05**	**1.22**	**2.24**

6.2 Influence of Mosaic Augmentation

Upon closer examination, we suspected that the use of mosaic augmentation [23] might be one of the reasons for it. This augmentation combines different portions of the original image to generate a transformed and augmented version, which can enhance the training dataset's diversity and variability.

We trained the model without employing this augmentation technique to evaluate its influence. The results can be depicted in Table 8. The baseline without mosaic achieves an average mAP of 34.74%, which is 7.83 lower than the mAP achieved by the baseline using mosaic augmentation. We performed more experiments without the mosaic augmentation on altitude, angle, and ensemble. These findings reinforce the significance of mosaic augmentation in this dataset. Mosaic augmentation aids the model in developing a more robust model, improving its performance. However, further tests should be conducted on datasets with orthogonal metadata to draw a more comprehensive understanding regarding the merits of one technique relative to the other.

Table 8. Average performance of each model without mosaic.

Model	mAP	mAP@50	mAP@75	AR1	AR10
Baseline	34.74	58.99	35.11	30.85	40.8
Single Domain with Altitude	35.55	60.75	36.27	31.67	42.10
Single Domain with Angle	35.97	60.88	36.94	32.08	42.30
Ensemble with Altitude and Angle	36.85	62.46	37.66	32.80	43.73

7 Conclusion

The use of UAVs for maritime surveillance is increasing, suggesting the importance of maritime object detection. To have a robust network capable of detecting objects in different altitudes and angles, we designed a robust network built on top of the YOLOv7, a state-of-the-art real-time object detector. The metadata was solely used during the training process, allowing the algorithm to be efficient at inference. We also designed an ensemble method combining multiple models, which is more complex and less suitable for deployment on edge devices.

Comprehensive experiments and analyses were conducted to evaluate this methodology's effectiveness. The gains achieved over the baseline were minimal when utilizing a single domain. An unexpected result was obtained when incorporating the adversarial domain alongside the other domain since the gains were lower than just a single domain. One potential reason for lower performance when utilizing multiple domains is that they may not be orthogonal regarding their impact on the captured images. As a result, during network training, the gradients generated during backpropagation may cancel each other. On the other hand, the ensemble model merging two models gave better results, having a higher improvement than the two previous methodologies. We also theorize that the mosaic augmentation would also increase the robustness of our neural network. We discovered that this augmentation helped the model to have a better performance, which is fundamental in this dataset. We also found that the gains using our approach without mosaic were higher, showing that mosaic helps the network become invariant to these variances.

Further experiments should be conducted using datasets that offer more diverse metadata. Also, to capture more real maritime surveillance scenarios, it is crucial to include representative conditions encompassing adverse weather and challenging sea conditions within the dataset, enabling the establishment of orthogonal domains. Unfortunately, publicly available datasets don't have images with adverse conditions along with metadata, as including them could provide further insights into the merits of this adversarial model training performance.

Acknowledgement. This work has been supported by Fundação para a Ciência e a Tecnologia (FCT) under the project UIDP/00048/2020.

References

1. Liang, X., Zhang, J., Zhuo, L., Li, Y., Tian, Q.: Small object detection in unmanned aerial vehicle images using feature fusion and scaling-based single shot detector with spatial context analysis. IEEE Trans. Circuits Syst. Video Technol. **30**(6), 1758–1770 (2020)
2. Ribeiro, R., Cruz, G., Matos, J., Bernardino, A.: A data set for airborne maritime surveillance environments. IEEE Trans. Circuits Syst. Video Technol. **29**(9), 2720–2732 (2019)
3. Sun, W., Gao, X.: Object detection in maritime scenarios based on SwinTransformer. In: Shmaliy, Y.S., Zekry, A.A. (eds.) 6th International Technical Conference on Advances in Computing, Control and Industrial Engineering (CCIE 2021). LNEE, pp. 786–798. Springer, Singapore (2022). https://doi.org/10.1007/978-981-19-3927-3_77
4. Chen, X., et al.: Video-based detection infrastructure enhancement for automated ship recognition and behavior analysis. J. Adv. Transp. **2020**, 1–12 (2020)
5. Varga, L.A., Kiefer, B., Messmer, M., Zell, A.: SeaDronesSee: a maritime benchmark for detecting humans in open water. In: 2022 IEEE/CVF Winter Conference on Applications of Computer Vision (WACV), pp. 3686–3696, Waikoloa, HI, USA. IEEE, January 2022

6. Kiefer, B., Ott, D., Zell, A.: Leveraging synthetic data in object detection on unmanned aerial vehicles, December 2021
7. Zhang, Y., Ge, H., Lin, Q., Zhang, M., Sun, Q.: Research of maritime object detection method in foggy environment based on improved model SRC-YOLO. Sensors **22**(20), 7786 (2022)
8. Jobson, D.J., Rahman, Z., Woodell, G.A.: Properties and performance of a center/surround retinex. IEEE Trans. Image Process. **6**(3), 451–462 (1997)
9. Nie, X., Yang, M., Liu, R.W.: Deep neural network-based robust ship detection under different weather conditions. In: 2019 IEEE Intelligent Transportation Systems Conference (ITSC), pp. 47–52, Auckland, New Zealand. IEEE, October 2019
10. Li, Q., Mou, L., Liu, Q., Wang, Y., Zhu, X.X.: HSF-Net: multiscale deep feature embedding for ship detection in optical remote sensing imagery. IEEE Trans. Geosci. Remote Sens. **56**(12), 7147–7161 (2018)
11. Chen, P., Li, Y., Zhou, H., Liu, B., Liu, P.: Detection of small ship objects using anchor boxes cluster and feature pyramid network model for SAR imagery. J. Mar. Sci. Eng. **8**(2), 112 (2020)
12. Zhang, Y., Guo, L., Wang, Z., Yang, Y., Liu, X., Fang, X.: Intelligent ship detection in remote sensing images based on multi-layer convolutional feature fusion. Remote Sens. **12**(20), 3316 (2020)
13. Hong, Z., et al.: Multi-scale ship detection from SAR and optical imagery via a more accurate YOLOv3. IEEE J. Sel. Topics Appl. Earth Observ. Remote Sens. **14**, 6083–6101 (2021)
14. Li, H., Deng, L., Yang, C., Liu, J., Gu, Z.: Enhanced YOLO v3 tiny network for real-time ship detection from visual image. IEEE Access **9**, 16692–16706 (2021)
15. Chen, L., Shi, W., Deng, D.: Improved YOLOv3 based on attention mechanism for fast and accurate ship detection in optical remote sensing images. Remote Sens. **13**(4), 660 (2021)
16. Li, Y., Yuan, H., Wang, Y., Zhang, B.: Maritime vessel detection and tracking under UAV vision. In: 2022 International Conference on Artificial Intelligence and Computer Information Technology (AICIT), pp. 1–4, September 2022
17. Li, Y., Yuan, H., Wang, Y., Xiao, C.: GGT-YOLO: a novel object detection algorithm for drone-based maritime cruising. Drones **6**(11), 335 (2022)
18. Farahani, A., Voghoei, S., Rasheed, K., Arabnia, H.R.: A brief review of domain adaptation, October 2020
19. Ganin, Y., Lempitsky, V.: Unsupervised domain adaptation by backpropagation, February 2015
20. Li, J., Xu, R., Ma, J., Zou, Q., Ma, J., Yu, H.: Domain Adaptive object detection for autonomous driving under foggy weather, October 2022
21. Wu, Z., Suresh, K., Narayanan, P., Xu, H., Kwon, H., Wang, Z.: Delving into robust object detection from unmanned aerial vehicles: a deep nuisance disentanglement approach. In: 2019 IEEE/CVF International Conference on Computer Vision (ICCV), pp. 1201–1210, Seoul, Korea (South). IEEE, October 2019
22. Ma, L., Meng, D., Huang, X., Zhao, S.: Vision-based formation control for an outdoor UAV swarm with hierarchical architecture. IEEE Access **11**, 75134–75151 (2023)
23. Kaur, P., Khehra, B.S., Mavi, Er.B.S.: Data augmentation for object detection: a review. In: 2021 IEEE International Midwest Symposium on Circuits and Systems (MWSCAS), pp. 537–543, August 2021

Streaming Graph-Based Supervoxel Computation Based on Dynamic Iterative Spanning Forest

Danielle Vieira[1] , Isabela Borlido Barcelos[1] , Felipe Belém[2] ,
Zenilton K. G. Patrocínio Jr.[1] , Alexandre X. Falcão[2] ,
and Silvio Jamil F. Guimarães[1(✉)]

[1] Laboratory of Image and Multimedia Data Science (ImScience), Pontifical Catholic University of Minas Gerais, Belo Horizonte, Brazil
{ddvieira,isabela.borlido}@sga.pucminas.br, {zenilton,sjamil}@pucminas.br
[2] Laboratory of Image Data Science (LIDS), University of Campinas, Campinas, Brazil
{afalcao,felipe.belem}@ic.unicamp.br

Abstract. Streaming video segmentation decreases processing time by creating supervoxels taking into account small parts of the video instead of using all video content. Thanks to the good performance of the Iterative Spanning Forest to compute Supervoxels (ISF2SVX) based on Dynamic Iterative Spanning Forest (DISF) for video segmentation framework we propose a new graph-based streaming video segmentation method for supervoxel generation by using dynamic iterative spanning forest framework, so-called StreamISF, based on a pipeline composed of six stages: (1) formation of the graph for each block of the video; (2) seed oversampling; (3) IFT-based supervoxel design; (4) reduction in the number of supervoxels; (5) spread of trees; and (6) creation of the segmented video. The difference in our proposed method is that it is unnecessary to have all the video in memory and the only previous information necessary to segment a block is the intersection frame between the blocks. Moreover, experimental results show that StreamISF creates supervoxels that maintain temporal coherence, producing very competitive measures compared to the state-of-the-art. Our code is publically available at https://github.com/IMScience-PPGINF-PucMinas/StreamISF.

Keywords: Graph-based method · Streaming · Supervoxel · Dynamic Iterative Spanning Forest

The authors thank the Pontifícia Universidade Católica de Minas Gerais – PUC-Minas, Coordenação de Aperfeiçoamento de Pessoal de Nível Superior – CAPES – (Grant COFECUB 88887.191730/2018-00, Grant PROAP 88887.842889/2023-00 – PUC/MG and Finance Code 001), the Conselho Nacional de Desenvolvimento Científico e Tecnológico – CNPq (Grants 303808/2018-7, 407242/2021-0, 306573/2022-9) and Fundação de Apoio à Pesquisa do Estado de Minas Gerais – FAPEMIG (Grant APQ-01079-23).

V. Vasconcelos et al. (Eds.): CIARP 2023, LNCS 14470, pp. 90–104, 2024.
https://doi.org/10.1007/978-3-031-49249-5_7

1   Introduction

In image and video applications, it is often essential to divide the image or video into meaningful regions or objects, in order to extract specific and relevant information from them. One may generate groups of connected elements (*i.e.*, superpixels or supervoxels) that share a common property (*e.g.*, color and texture). When numerous groups are generated, the object can be effectively defined by its comprising regions, being the major premise of superpixel and supervoxel segmentation algorithms. Such methods are applied to many contexts such as: (i) object detection [11,13]; (ii) cloud connectivity [14,18,20]; and (iii) long-range tracking [15].

In general, three properties are desirable in video supervoxel segmentation: (i) spatiotemporal boundary adherence; (ii) computational efficiency; and (iii) ability to control the number of supervoxels generated. However, no supervoxel segmentation algorithm has all these characteristics [19]. For early video processing, one can interpret it as a three-dimensional spatiotemporal volume and segment its objects. Streaming video segmentation refers to the process of dividing a continuous video stream into smaller segments in order to reduce memory consumption and improve segmentation processing time. Usually, good early video segmentation methods preserve temporal coherence. However, this is not the case when video segmentation streaming is applied. Thus, one of the major challenges for streaming video segmentation is to ensure its temporal coherence.

The *Graph-based supervoxel* (GB) [10] is an image segmentation method based on graphs with good boundary adherence but is computationally expensive. The *hierarchical GB* (GBH) [12] considers the GB strategy for computing a hierarchical iterative method; while the *stream GBH* (sGBH) [21] extends the latter for online video segmentation. In [7,8], the authors proposed a hierarchical segmentation method – HOScale (Hierarchical video segmentation using an Observation Scale) – based on the same criterion of [10] that removes the need for parameter tuning and for the computation of video segmentation at finer levels. In their proposal, the video segmentation strategy is not dependent on the hierarchical level, and consequently, it is possible to compute any level without the previous ones. Therefore, the time for computing a segmentation is almost the same for any specified level. Moreover, according to experimental results presented in [8], HOScale produces good quantitative and qualitative results when compared to other methods.

Following [21], StreamHOScale [16] divides a video into frame blocks and transforms the streaming video segmentation into a graph partitioning problem for each frame block. StreamHOScale merges the segmentation blocks using a simple and efficient strategy to achieve temporal coherence. Our method also performs segmentation in blocks and does not need the entire video in memory, as does StreamHOScale, but uses an IFT-based supervoxel design for segmentation. However, due to its hierarchical property, wrong borders computed in the lowest levels could be persistent for the highest ones. This is why we have decided to use Dynamic Iterative Spanning Forest (DISF). Even this one is not a hierarchical method, it starts from a very large number of seeds, and consequently, regions to a desired number of regions by using a strategy of removing seeds allowing, then, a competition between them for correcting wrong borders.

In [2], a supervoxel segmentation framework for video segmentation, named *Iterative Spanning Forest to compute Supervoxels* (ISF2SVX), was proposed. This method was inspired by the *Iterative Spanning Forest* (ISF) [17], which is a recent superpixel segmentation framework. This method could be seen as an application of the DISF for supervoxel computation in which the graph is obtained by the video instead. Similar to ISF, the proposed approach was composed of independent steps: (i) graph construction; (ii) seed sampling; (iii) supervoxel generation; and (iv) seed recomputation. In step (i), the video volume is converted to a directed graph representation which will be used as input for determining the seeds in step (ii). Then, for several iterations, ISF2SVX generates supervoxels through the *Image Foresting Transform* (IFT) [9] using improved seed sets—in steps (iii) and (iv), respectively.

Inspired by [16], we propose, in this article, a promising strategy for streaming the ISF2SVX method that does not need all the video in memory, its processing time is 30% faster and the only previous information needed to segment a block is the intersection frame.

Figure 1 illustrates examples of results obtained by ISF2SVX and the proposed method, so-called StreamISF, for 10 and 500 supervoxels. It is possible to observe in Fig. 1(a) that in the first segmented blocks, the skater's dress is a region that is not in the ISF2SVX result, and even after the dress does not appear in the last blocks due to tree mergers during the propagation of the labels of the intersection trees, the sign on the wall is more detailed in our proposed method in all the images of Fig. 1(a) and 1(b), both for 10 supervoxels and for 500 supervoxels.

This paper is organized as follows. In Sect. 2, important concepts used in this work such as graphs and IFT are clarified. In Sect. 3, the methodology for the proposed streaming segmentation approach is explained. In Sect. 4, we describe the experiments carried out, the generation of metrics, and a comparison with other segmentation methods. And in Sect. 5 the final considerations and future works are presented.

2   Theoretical Background

To facilitate the understanding of our method, we explain the necessary concepts and techniques related to our proposal. We first introduce some graph notions to present the core delineation algorithm of our proposal: *Image Foresting Transform* (IFT) [9] framework.

2.1   Graph

A *video* V can be represented as a pair $\mathsf{V} = (\mathcal{V}, \mathbf{I})$ in which $\mathcal{V} \subseteq \mathbb{N}^3$ denotes the set of *volume elements* (*i.e.*, voxels), and $\mathbf{I}$ maps every $v \in \mathcal{V}$ to a feature vector $\mathbf{I}(v) \in \mathbb{R}^m$. One can see that, for $m = 3$, V is a colored video (*e.g.*, RGB

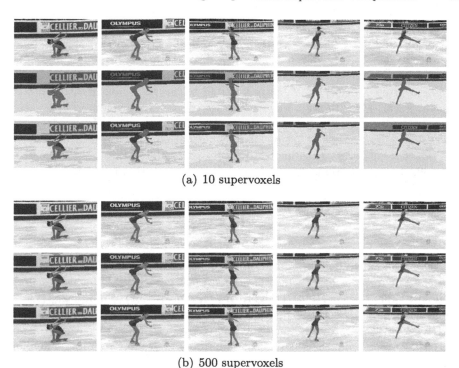

(a) 10 supervoxels

(b) 500 supervoxels

Fig. 1. Examples of video segmentations for a video extracted of the GATech. The original frames are illustrated in the first row. We illustrate examples of the proposed method changing the seed sampling. We illustrate results for (a) 10 and (b) 500 super-voxels, considering ISF2SVX-GRID-DYN and StreamISF with $k = 30\%$ and allowing tree merges (second and third rows). Each resulting region is colored by its mean color.

or CIELAB colorspaces). One may create a *simple graph* (*i.e.*, no loops and no parallel edges) $\mathsf{G} = (\mathcal{N}, \mathcal{E})$, derived from V, in which $\mathcal{N} \subseteq \mathcal{V}$ denotes the *vertex* set and $\mathcal{E} \subset \mathcal{N}^2$, the *edge* set. Two nodes $v_i, v_j \in \mathcal{N}$ are said to be *adjacent* if $(v_i, v_j) \in \mathcal{E}$. From that, a weighted-vertex graph $(\mathsf{G}, \mathbf{I})$ may be defined in which $\mathbf{I}(v) \in \mathbb{R}^m$. In this work, the elements in $\mathcal{E}$ are oriented *arcs* (*i.e.*, G is a *digraph*). Consider $\pi_{s \rightsquigarrow t} = \langle s = v_1, v_2, \ldots, v_n = t \rangle$ to be a finite sequence of adjacent nodes (*i.e.*, a *path*) in which $(v_i, v_{i+1}) \in \mathcal{E}$ for $1 \leq i < n$. For simplicity, we may omit the path *origin* voxel by writing π_t. For $n = 1$, $\pi_t = \langle t \rangle$ is said to be *trivial*. We denote the *extension* of a path π_s by an arc $(s, t) \in \mathcal{E}$ as $\pi_s \cdot \langle s, t \rangle$ with the two instances of s being merged into one.

2.2 Image Foresting Transform

The *Image Foresting Transform* (IFT) [9] is a framework for the development of image processing operators based on connectivity and has been used to reduce image processing tasks as optimum-path forest computations over the image graph. As indicated by the authors [9], the IFT is independent of the input's dimensions and, therefore, the relation between pixels (or voxels) in such dimensionality can effectively be represented by an *adjacency relation* between them. In this work, we consider the IFT version restricted to a *seed set* $\mathcal{S} \subset \mathcal{N}$.

For a given arc $(s,t) \in \mathcal{E}$, it is possible to assign a non-negative *arc-cost* value $\mathbf{w}_*(s,t) \in \mathbb{R}^+$ through an *arc-cost function* $\mathbf{w}_*$. A common approach is to compute the ℓ_2-norm between the nodes' features—*i.e.*, $\|\mathbf{I}(s) - \mathbf{I}(t)\|_2$ for $s,t \in \mathcal{N}$. Consider Π_G the set of all possible paths in G. Then, a *connectivity function* $\mathbf{f}_*$ maps every path in Π_G to a *path-cost value* $\mathbf{f}_*(\pi_t) \in \mathbb{R}^+$. One of the most effective connectivity functions for object delineation is the $\mathbf{f}_{\max}$ function:

$$\mathbf{f}_{\max}(\langle t \rangle) = \begin{cases} 0 & \text{if } t \in \mathcal{S}, \\ +\infty & \text{otherwise} \end{cases} \tag{1}$$

$$\mathbf{f}_{\max}(\pi_s \cdot \langle s,t \rangle) = \max\{\mathbf{f}_{\max}(\pi_s), \mathbf{w}_*(s,t)\}$$

A path π_t^* is said to be *optimum* if, for any other path $\tau_t \in \Pi_G$, $\mathbf{f}_*(\pi_t^*) \leq \mathbf{f}_*(\tau_t)$.

Let $\mathbf{C}$ be a *cost map* in which assigns, to every path $\pi_t \in \Pi_G$, its respective path-cost value $\mathbf{f}_*(\pi_t)$. The IFT algorithm minimizes $\mathbf{C}(t) = \min_{\forall \pi_t \in \Pi_G}\{\mathbf{f}_*(\pi_t)\}$ whenever $\mathbf{f}_*$ satisfies certain conditions [6]. First, the IFT assigns path-costs to all trivial paths accordingly and, then, it computes optimum paths in a non-decreasing order, from the seeds to the remaining nodes in the graph. Therefore, independently if $\mathbf{f}_*$ suffices the desired properties in [6], the IFT always generates a spanning forest and, consequently, each supervoxel is a unique tree. During the segmentation process, a *predecessor map* $\mathbf{P}$ is generated and defined. Such map assigns any node $t \in \mathcal{N}$ to its *predecessor* s in the optimum path $\pi_s^* \cdot \langle s,t \rangle$, or to a distinctive marker $nil \notin \mathcal{N}$—in such case, t is said to be a *root* of $\mathbf{P}$. In this work, every seed is a root of $\mathbf{P}$. One may see that $\mathbf{P}$ is a representation of an *optimum-path forest*, and it allows to recursively obtain the optimum-path root $\mathbf{R}(t)$ of t and its root's label $\mathbf{L}(\mathbf{R}(t))$.

3   A Strategy for Supervoxel Computation Based on Dynamic Iterative Spanning Forest

In this work, we propose an approach for supervoxel computation based on *Iterative Spanning Forest* (ISF) [17] superpixel framework which was successfully applied to video segmentation in [2], and that it is not necessary to have the entire video in memory, with only the intersection image as prior information for block segmentation. Our proposal, so-called StreamISF, adopts a six-step methodology: (1) creation of block graphs; (2) seed sampling; (3) IFT-based supervoxel delineation; (4) seed set recomputation; (5) propagation of label trees; and (6) video segmented. This proposed pipeline is illustrated in Fig. 2. It is important to note that this work performs video segmentation in blocks followed by a propagation of supervoxel trees computed by the ISF-based *Dynamic and Iterative Spanning Forest* (DISF) [3] method. Thus, StreamISF could be seen as an application of the DISF for supervoxel computation in which the graph is obtained by the video.

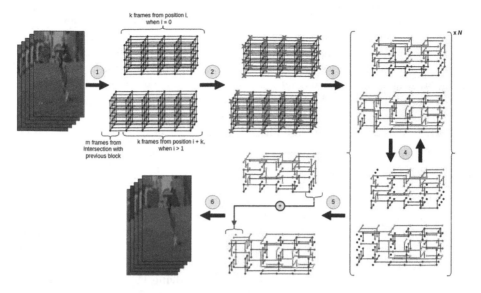

Fig. 2. Outline for streaming graph-based supervoxel segmentation based on iterative spanning forest. Our method is composed of six stages: (1) formation of the graph for each block of the video; (2) seed oversampling; (3) IFT-based supervoxel design; (4) reduction in the number of supervoxels; (5) spread of trees; and (6) creation of the segmented video.

3.1 Video Block Computation and Seed Sampling

First, the video is divided into k-sized blocks, in which the first block has k frames and the remaining ones have $(k + 1)$. Therefore, each block, except the first one, has an intersection frame with the previous block. The intersection frame is used to try to maintain the temporal coherence of the video, which is based only on the color information. Let a video block $V_i \in V$, in which $\langle V_1, \ldots, V_n \rangle$ is a sequence of n blocks in temporal ordering, one may compute a graph G_i for each V_i. Similar to [3], in StreamISF, a seed oversampling is performed at each G_i using a grid sampling scheme [1] (hereinafter named GRID), which selects N_0 equally distanced seeds within the graph.

3.2 Supervoxel Generation

Once seeds are sampled, the supervoxels are generated using the IFT algorithm considering a connectivity function $\mathbf{f}_*$ and an arc-cost function $\mathbf{w}_*$. In this work, we consider the $\mathbf{f}_{\max}$ connectivity function for computing the path-costs.

In [17], the authors recall an arc-cost function $\mathbf{w}_1(p, q) = (\alpha \|\mathbf{I}(\mathbf{R}(p)) - \mathbf{I}(q)\|_2)^\beta + \|p - q\|_2$ in which $\alpha \in \mathbb{R}^+_*$ permits the user to control the regularity of the superpixels and to control their adherence to boundaries through a factor $\beta \in \mathbb{R}^+_*$. However, superpixel and supervoxel regularity tend to prejudice the object delineation performance [3].

In DISF, the arc-costs are computed dynamically considering mid-level super-pixel features, using a function first proposed in [4]. Let a graph $G_i = (\mathcal{N}, \mathcal{E})$ be a simple graph computed on the block i and $(G_i, \mathbf{I})$ be the associated weighted-vertex graph to the block i. Let $\mathcal{T}_x \subset \mathcal{N}$ be an optimum-path *growing* tree rooted in a node $x \in \mathcal{N}$, and let $\mu(\mathcal{T}_x)$ be its mean feature vector. Then, the arc-cost function $\mathbf{w}_2$ can be formally defined as $\mathbf{w}_2(p,q) = \|\mu(\mathcal{T}_{\mathbf{R}(p)}) - \mathbf{I}(q)\|_2$. The function $\mathbf{w}_2$ has proven to be more effective than classic arc-cost functions for both superpixel segmentation [3] and for interactive image segmentation [4].

3.3   Seed Recomputation

Like ISF2SVX, the fourth step in StreamISF iteratively updates the seed set $\mathcal{S}$ to improve the supervoxel delineation for subsequent iterations. This update can be performed by including, shifting, or removing the seeds in $\mathcal{S}$. Similar to other ISF-based methods [2,3], after seed sampling, we iteratively perform supervoxel generation followed by seed recomputation. This recomputation procedure promotes the growth of relevant supervoxels, by removing the irrelevant ones and maintaining the competition among the primers. Let $N_f^i \in \mathbb{N}$ be the desired number of final supervoxels per block, in which $N_0 \gg N_f^i$ and $1 \leq i \leq b$. At each iteration $j \in \mathbb{N}$, $\mathbf{M}(j) = \max\{N_0 \exp^{-j}, N_f^i\}$ relevant seeds are maintained for the subsequent iteration $j + 1$, while the remaining ones are discarded. The stopping criterion is reaching the desired number of supervoxels, which is often less than 10—a common value for many iterative methods.

The $\mathbf{M}(j)$ relevant seeds may be selected by a combination of their sizes and contrast [3] in which the former indicates the supervoxel's growth ability, and the latter, whether the supervoxel is located in a homogeneous region (thus, probably irrelevant). Let $\mathcal{B}$ be a *tree adjacency relation*, which defines the immediate neighbors of any supervoxel. Then, with the use of a priority queue, a relevance of a seed s can be measured by a function $\mathbf{V}(s) = \frac{|\mathcal{T}_s|}{|\mathcal{N}|} \min_{\forall (\mathcal{T}_s, \mathcal{T}_r) \in \mathcal{B}} \{\|\mu(\mathcal{T}_s) - \mu(\mathcal{T}_r)\|_2\}$, which $\mathcal{T}_r$ is an *adjacent supervoxel* of $\mathcal{T}_s$. It is important to observe that each supervoxel is related to the tree, so-called supervoxel tree.

3.4   Propagating Labels of the Supervoxel Trees

Our proposal allows for computing a segmentation in a video block without having to keep the previous blocks in memory. Also, StreamISF tries to preserve temporal coherence with a minimum amount of information, which is, in this case, one frame-sized. After computing supervoxels for a graph G_i (related to the ith video block), each supervoxel tree is associated to labels in order to facilitate the propagation of the information between different blocks, thus we propagate the labels from the previous block to G_i based on the intersection frame (first frame) in G_i and the labels of the last frame of the previous block. To propagate the labels from one video block to the next, we propose two strategies: (i) merging trees to improve supervoxel coherence, or (ii) prioritizing the highest number of supervoxels. An example of each strategy is illustrated in Fig. 3.

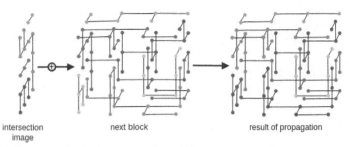

intersection   next block   result of propagation
image

(a) First method of propagation: Allows supervoxel tree merges.

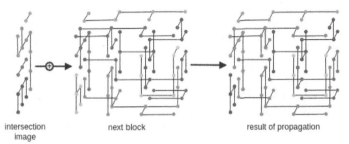

intersection   next block   result of propagation
image

(b) Second method of propagation: Does not allow merging of supervoxel trees.

Fig. 3. Example of propagation methods.

Let $\mathsf{G}_i = (\mathcal{N}_i, \mathcal{E}_i)$ and $\mathsf{G}_i^j = (\mathcal{N}_i^j, \mathcal{E}_i^j)$ be the simple graphs for the block i and for the frame j of the block i. Let $(\mathsf{G}_i, \mathbf{I})$ be the weighted-vertex graph for the ith video block and $(\mathsf{G}_i^j, \mathbf{I}^j)$ a subgraph containing vertices related with its jth frame. One may represent $\mathcal{N}_i^j$ as a set of disjoint subsets $\mathcal{L}_i^j = \langle \tau_1, ..., \tau_l \rangle$, in which τ_t contains the vertices of the same tree in G_i, and $\bigcup_{\tau_t \in \mathcal{L}_i^j} \tau_t = \mathcal{N}_i^j$. Let the ith video block with k frames and the sets $\tau_p \in \mathcal{L}_{i-1}^k$, and $\tau_q \in \mathcal{L}_i^1$. In tree label propagation across different frames, the following cases may require label propagation: (1) $\tau_q = \tau_p$; and (2) $\tau_p \cap \tau_q \neq \{\emptyset\}$, but $\tau_p \neq \tau_q$.

In both propagation strategies, when $\tau_p = \tau_q$ the label of τ_p is propagated to τ_q (e.g., the orange tree being propagated to the light blue supervoxel tree in Fig. 3(a)). When $\tau_p \cap \tau_q \neq \{\emptyset\}$, but $\tau_q \neq \tau_p$, one may have a tree competition between τ_p and $\tau_r \in \mathcal{L}_{i-1}^k$ or τ_q and $\tau_r \in \mathcal{L}_i^1$. For the former, at least two trees in the intersection compete for a tree in the block. In this case, we propagate the label of $\tau = \mathrm{argmax}_{\tau_x \in \mathcal{L}_{i-1}^k} \{ |\tau_x \cap \tau_q| \}$ (e.g., in Fig. 3(a), the green and pink trees, at the intersection frame, competing for the green tree of the next block). When a tie occurs (i.e., when $|\tau_q \cap \tau_q| = |\tau_r \cap \tau_q|$), we propagate the label of $\tau = \mathrm{argmax}_{\tau_x \in \mathcal{L}_i^k}\{|\tau_x|\}$. Finally, when a tree in the intersection τ_p may propagate its label for at least two trees in the block, we may perform one of two strategies: (i) propagate the label in τ_p to the trees at the block, merging them (merging propagation strategy); or (ii) propagate the label in τ_p to the tree $\tau =$

$\mathrm{argmax}_{\tau_x \in \mathcal{L}_i^1}\{|\tau_x|\}$ and maintain the label of the others (non-merging strategy). Our first propagation strategy prioritizes time coherence by propagating trees even when they are not exactly the same (*i.e.*, when $\tau_p \cap \tau_q \neq \{\emptyset\}$, but $\tau_p \neq \tau_q$). Therefore, it may perform tree merging, reducing the number of supervoxels, which hampers maintaining the exact final number of supervoxels. For instance, in Fig. 3(a), the purple tree is propagated to the lighter pink and lilac trees in the block, causing these two trees to merge, and reducing the number of final supervoxels in one. Conversely, the second propagation strategy prioritizes the desired number of final supervoxels, choosing a unique tree to propagate the label and maintaining the others. In Fig. 3(b), instead of propagating the purple tree in the intersection frame to both trees and merging them, it propagates to one.

The final number of supervoxels depends on the number of supervoxels produced at each block, which may be greater or lesser than the desired and directly depends on the N_f^i for each block i. Therefore, we assign values to N_f^i according to the label propagation strategy used. Let N_f^{max} be the final number of supervoxels desired, and γ and δ are the decreasing factors for each propagation strategy. For the merging-based strategy, we first set $N_f^1 \gg N_f^{max}$, the decreasing factor $\gamma(i) = (N_f^{max} \times (i-1))/100$, and $N_{min} = \rho * N_f^{max}$ as minimum number of supervoxels at any block. Then, for each block $i > 1$, we set $N_f^i = \max\{N_f^{i-1} - \gamma, N_{min}\}$, preventing the graphs of video blocks from generating a number of supervoxels lesser than N_{min}. On the other hand, for the non-merging strategy, let κ_i the number of supervoxels in G_{i-1} that are not into G_{i-2}, and $\delta_{i-1} = N_f^{i-1} + \kappa_{i-1}$. Then, for each block $i > 1$, we set $N_f^i = \frac{N_f^{max} - \delta_{i-1}}{b-i}$. In contrast to the merging-based strategy, the non-merging one prevents the graphs from generating a high number of supervoxels.

4   Experimental Analysis

In this work, we propose a strategy for streaming graph-based supervoxel computation, named StreamISF, in which the size of each block is flexible and it depends on the size of the video. Here, we consider the size of each frame-block equal to 10%, 20%, and 30% of the video size. We also made some experiments when the number of frames for each block is independent of the video's size, for instance, 10 and 20 frames. It is important to observe that when $k = 100\%$, this method becomes the ISF2SVX. We compared our approaches with different state-of-the-art methods: (a) ISF2SVX; (b) StreamGBH; and (c) StreamHOScale. The number of supervoxels varied from 200 to 800 and, for the baselines, the recommended parameter settings were used. StreamISF$_w$ is when propagating intersection labels allows tree merges and StreamISF$_n$ does not allow merges.

We evaluated on Chen [5] dataset, some samples can be seen in Fig. 4, which is a subset of the well-known xiph.org videos that have been supplemented with a 24-class semantic pixel labeling set (the same classes from the MSRC object-segmentation dataset). The eight videos in this set are densely labeled with

Fig. 4. Example of images from Chen dataset.

semantic pixels and have an average of 85 fpv, minimum 69 fpv, and maximum 86 fpv, leading to a total of 639 annotated frames. This dataset allows us to evaluate the supervoxel methods against human perception. The annotated frames correspond to semantic pixels, thus objects spatially disconnected have the same label, thereby the evaluations using these annotated frames do not over an accurate label placement of supervoxels in the videos.

In terms of measures, we selected five classic evaluation metrics: (a) 3D boundary recall (BR); (b) 3D undersegmentation error (UE); (c) Explained variation (EV); and (d) Mean duration. BR measures the quality of the spatiotemporal boundary delineation. UE calculates the fraction of object supervoxels overlapping background voxels—and vice-versa—($i.e.$, lower is better). EV measures the method's ability to describe the video's color variations through its supervoxels ($i.e.$, higher is better). Finally, the mean supervoxel duration measures if a supervoxel perpetuates throughout the frames, indicating a temporal coherence to the object which it compounds ($i.e.$, higher is better).

4.1 Quantitative Analysis

Figures 5 and 6 present the quantitative results. Considering UE on Figs. 5, one may see that StreamISF, with 26-neighborhood and k equal to 20% and 30% have competitive or better results than the compared works, with significantly less supervoxel leakage than StreamHOScale. Similarly, StreamISF has competitive BR results, in which StreamISF with 26-neighborhood with k equal to 20% and 30% have a higher BR than StreamGBH. As stated in [3,17] path-based methods, and more specifically IFT-based methods, are known to be effective solutions in object delineation, thus justifying why the ISF2SVX method in its variations have the best boundary recall results.

As one may see in Fig. 6, since IFT minimizes the cumulative path cost, the internal variation of the supervoxels tends to be significantly reduced. Thus, compared to other streaming video segmentation, the StramISF variations achieved the best EV results. Maintaining temporal coherence is a challenging task for streaming-based methods since they intend to reduce memory consumption and allow for real-time video processing. The mean duration tries to capture the temporal coherence of the supervoxels, and our method manages to maintain the

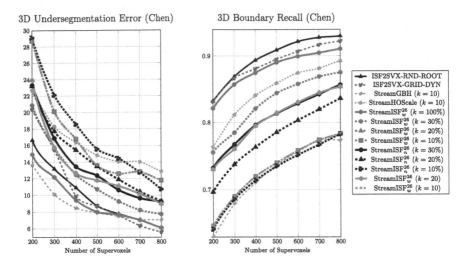

Fig. 5. A comparison between our method StreamISF, and the methods StreamGBH ($k = 10$), StreamHOScaleand ISF2SVX when applied to Chen datasets. The comparison is based on the following metrics: (i) 3D undersegmentation error; and (ii) 3D boundary recall .

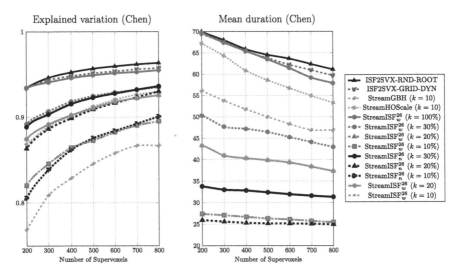

Fig. 6. A comparison between our method StreamISF, and the methods StreamGBH ($k = 10$), StreamHOScaleand ISF2SVX when applied to Chen datasets. The comparison is based on the explained variation and mean duration.

temporal coherence between the blocks, but as our method is still not able to guarantee the final number of supervoxels in the video, which is an approximate number, our method in this metric obtains lower results than other segmentation methods. StreamISF with 26-neighborhood with k equal to 30% is the one with

Table 1. Video execution time of the images in Fig. 7 from the Chen dataset. The column k indicates the block size (in frames) in relation to the video size. In both StreamISF and ISF2SVX the parameters were 5000 initial seeds and 100 final supervoxels.

Method	k	Tree merging	Time (s)
ISF2SVX-GRID-DYN	–	–	34.368
ISF2SVX-RND-ROOT	–	–	31.245
StreamISF-GRID-DYN	30%	✗	30.478
StreamISF-RND-ROOT	30%	✗	27.062
StreamISF-GRID-DYN	20%	✗	29.470
StreamISF-RND-ROOT	20%	✗	26.120
StreamISF-GRID-DYN	10%	✗	30.795
StreamISF-RND-ROOT	10%	✗	29.181
StreamISF-GRID-DYN	30%	✓	25.656
StreamISF-RND-ROOT	**30%**	✓	**22.334**
StreamISF-GRID-DYN	20%	✓	25.746
StreamISF-RND-ROOT	**20%**	✓	**22.756**
StreamISF-GRID-DYN	10%	✓	24.590
StreamISF-RND-ROOT	**10%**	✓	**22.190**

the best temporal coherence among the variations of our own method, according to Fig. 6.

According to Table 1, when comparing the average execution time of Stream-ISF with ISF2SVX, our proposal can improve efficiency and reduce memory consumption. As you can see, the efficiency of StreamISF improves by around 30% by allowing tree merging and using a random sampling strategy. In Stream-ISF, when two or more trees can receive the label of the same tree from the intersection frame, performing tree merging is a more efficient task (in terms of execution time) than choosing which tree to propagate to. Furthermore, when sampling seeds in random positions, there is a reduction in computational cost in relation to grid sampling, since grid sampling requires the computation of equally spaced positions.

4.2   Qualitative Analysis

In Fig. 7, we compare the StreamISF variant with 26-neighborhood and $k = 30\%$ with other segmentation methods. As can be seen, our approach manages to generate large supervoxels in non-significant regions (*e.g.*, the grass), while effectively delineating even small important regions (*e.g.*, the player's head and the letters on the jersey). In contrast, due to the fusions in the propagation of the labels of the intersection trees, some of the temporal coherence of the colors is lost in the course of the video.

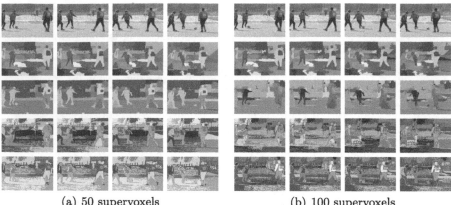

| (a) 50 supervoxels | (b) 100 supervoxels |

Fig. 7. Example extracted from Chen dataset. The first row contains the original frames, and the following rows, from top to bottom have results with 50 and 100 supervoxels obtained from StreamGBH ($k = 10$), StreamHOScale ($k = 10$), StreamISF (26-neighborhood e $k = 30\%$), and ISF2SVX-GRID-DYN.

5   Conclusions and Future Work

In this paper, we propose a new supervoxel segmentation streaming framework, named *Stream Iterative Spanning Forest* (StreamISF), which was inspired by [16] and the *Iterative Spanning Forest for Supervoxels* (ISF2SVX), supervoxel segmentation framework. Our method allows video segmentation without high memory consumption, since only the video block and its intersection image are stored, and not the whole video as is done in the ISF2SVX method. Also, according to the presented results, our algorithm is better than some compared methods that also perform streaming video segmentation.

For future works, we intend to study the performance of StreamISF considering new ways of propagating supervoxel tree labels to improve the supervoxels' delineation with fewer tree merges and improving temporal coherence. Also, we intend to study strategies to ensure the number of final supervoxels.

References

1. Achanta, R., Shaji, A., Smith, K., Lucchi, A., Fua, P., Süsstrunk, S.: Slic superpixels compared to state-of-the-art superpixel methods. Trans. Pattern Anal. Mach. Intell. **34**(11), 2274–2282 (2012)
2. Jerônimo, C., et al.: Graph-based supervoxel computation from iterative spanning forest. In: Lindblad, J., Malmberg, F., Sladoje, N. (eds.) DGMM 2021. LNCS, vol. 12708, pp. 404–415. Springer, Cham (2021). https://doi.org/10.1007/978-3-030-76657-3_29

3. Belém, F., Guimarães, S., Falcão, A.: Superpixel segmentation using dynamic and iterative spanning forest. Signal Process. Lett. **27**, 1440–1444 (2020)
4. Bragantini, J., Martins, S.B., Castelo-Fernandez, C., Falcão, A.X.: Graph-based image segmentation using dynamic trees. In: Vera-Rodriguez, R., Fierrez, J., Morales, A. (eds.) CIARP 2018. LNCS, vol. 11401, pp. 470–478. Springer, Cham (2019). https://doi.org/10.1007/978-3-030-13469-3_55
5. Chen, A., Corso, J.: Propagating multi-class pixel labels throughout video frames. In: Western New York Image Processing Workshop, pp. 14–17 (2010)
6. Ciesielski, C., Falcão, A., Miranda, P.: Path-value functions for which Dijkstra's algorithm returns optimal mapping. J. Math. Imaging Vision **60**(7), 1025–1036 (2018)
7. De Souza, K.J.F., et al.: Hierarchical video segmentation using an observation scale. In: Hirata, N., Nedel, L., Silva, C., Boyer, K. (eds.) SIBGRAPI 2013 (XXV Conference on Graphics, Patterns and Images), Arequipa (Aug 2013)
8. De Souza, K.J.F., De Albuquerque Araújo, A., Do Patrocínio, Jr., Z.K.G., Guimarães, S.J.F.: Graph-based hierarchical video segmentation based on a simple dissimilarity measure. Pattern Recogn. Lett. **47**, 85–92 (2014). https://doi.org/10.1016/j.patrec.2014.02.016
9. Falcão, A., Stolfi, J., Lotufo, R.: The image foresting transform: theory, algorithms, and applications. Trans. Pattern Anal. Mach. Intell. **26**(1), 19–29 (2004)
10. Felzenszwalb, P., Huttenlocher, D.: Efficient graph-based image segmentation. Int. J. Comput. Vision **59**(2), 167–181 (2004)
11. Griffin, B.A., Corso, J.J.: Video object segmentation using supervoxel-based gerrymandering. arXiv preprint arXiv:1704.05165 (2017)
12. Grundmann, M., Kwatra, V., Han, M., Essa, I.: Efficient hierarchical graph-based video segmentation. In: Computer Vision and Pattern Recognition (CVPR), pp. 2141–2148. IEEE (2010)
13. Oneata, D., Revaud, J., Verbeek, J., Schmid, C.: Spatio-temporal object detection proposals. In: Fleet, D., Pajdla, T., Schiele, B., Tuytelaars, T. (eds.) ECCV 2014. LNCS, vol. 8691, pp. 737–752. Springer, Cham (2014). https://doi.org/10.1007/978-3-319-10578-9_48
14. Papon, J., Abramov, A., Schoeler, M., Worgotter, F.: Voxel cloud connectivity segmentation-supervoxels for point clouds. In: Proceedings of the IEEE Conference on Computer Vision and Pattern Recognition, pp. 2027–2034 (2013)
15. Sheng, H., Zhang, X., Zhang, Y., Wu, Y., Chen, J., Xiong, Z.: Enhanced association with supervoxels in multiple hypothesis tracking. IEEE Access **7**, 2107–2117 (2018)
16. de Souza, K.J.F., de Albuquerque Araújo, A., Guimarães, S.J.F., do Patrocínio, Z.K.G., Cord, M.: Streaming graph-based hierarchical video segmentation by a simple label propagation. In: 28th SIBGRAPI Conference on Graphics, Patterns and Images (SIBGRAPI 2015), Salvador, 26–29 August 2015, pp. 119–125. IEEE Computer Society (2015). https://doi.org/10.1109/SIBGRAPI.2015.23
17. Vargas-Muñoz, J., Chowdhury, A., Alexandre, E., Galvão, F., Miranda, P., Falcão, A.: An iterative spanning forest framework for superpixel segmentation. Trans. Image Process. **28**(7), 3477–3489 (2019)
18. Verdoja, F., Thomas, D., Sugimoto, A.: Fast 3D point cloud segmentation using supervoxels with geometry and color for 3D scene understanding. In: 2017 IEEE International Conference on Multimedia and Expo (ICME), pp. 1285–1290. IEEE (2017)
19. Wang, B., et al.: Real-time hierarchical supervoxel segmentation via a minimum spanning tree. Trans. Image Process. **29**, 9665–9677 (2020)

20. Wu, F., et al.: Rapid localization and extraction of street light poles in mobile lidar point clouds: a supervoxel-based approach. IEEE Trans. Intell. Transp. Syst. **18**(2), 292–305 (2016)
21. Xu, C., Xiong, C., Corso, J.J.: Streaming hierarchical video segmentation. In: Fitzgibbon, A., Lazebnik, S., Perona, P., Sato, Y., Schmid, C. (eds.) ECCV 2012. LNCS, vol. 7577, pp. 626–639. Springer, Heidelberg (2012). https://doi.org/10.1007/978-3-642-33783-3_45

Improving Pest Detection via Transfer Learning

Dinis Costa[1]($\boxtimes$)[iD], Catarina Silva[1][iD], Joana Costa[1,2][iD],
and Bernardete Ribeiro[1][iD]

[1] CISUC, Department of Informatics Engineering, Coimbra, Portugal
ddcosta@student.dei.uc.pt, {catarina,joanamc,bribeiro}@dei.uc.pt
[2] Polytechnic Institute of Leiria, School of Technology and Management, Leiria,
Portugal

Abstract. Pest monitoring models play a vital role in enabling informed decisions for pest control and effective management strategies. In the context of smart farming, various approaches have been developed, surpassing traditional techniques in both efficiency and accuracy. However, the application of Few-Shot Learning (FSL) methods in this domain remains limited. In this study, we aim to bridge this gap by leveraging Transfer Learning (TL). Our findings highlight the considerable efficacy of TL techniques in this context, showcasing a significant 24% improvement in mAP performance and a 10% reduction in training time, thereby enhancing the efficiency of the model training process.

Keywords: Machine Learning · Transfer Learning · Few-shot Learning · Pest Detection · Object Detection

1 Introduction

In the agriculture industry, several algorithms have been employed alongside Internet of Things (IoT) technologies to enhance productivity [9]. Some of these smart techniques are specifically tailored to enhance plant growth, such as precision trimming methods that promote uniform growth across the plantation, leading to consistent and similar fruits. Additionally, some techniques utilize available data on weather and soil conditions to aid farmers in planning optimal planting schedules, forecasting crop yields, and estimating food requirements. A particular application with significant potential is pest detection, as it can effectively enhance crop yields and minimize pesticide usage by detecting pests at early stages. Traditional pest detection methods rely on expert technicians manually inspecting the plantation to identify and count the pests [1]. Since this task is challenging due to the difficulty of identifying some pests, and also time-consuming and error-prone, the number of pests in a plantation is often extrapolated from the number of pests counted by the experts in a particular area of the plantation.

Smart methods have the advantage of surpassing traditional approaches in pest detection, as they do not always necessitate the presence of an expert technician and can efficiently cover larger regions of a plantation. However, training

© Springer Nature Switzerland AG 2024
V. Vasconcelos et al. (Eds.): CIARP 2023, LNCS 14470, pp. 105–116, 2024.
https://doi.org/10.1007/978-3-031-49249-5_8

such models requires high-quality labeled data, which, unfortunately, is scarce in this domain. The limited availability of labeled data poses a significant challenge, leading to the development of innovative methods that aim to reduce the reliance on large volumes of real data while effectively addressing these constraints.

With that in mind, there is a growing demand for lighter, less data-intensive, and energy-efficient AI models. Optimized learning processes are required to achieve data-efficient AI, ensuring reduced input requirements without compromising the quality of the output. As an example, Active Learning (AL) seeks to alleviate the requirement for extensive datasets to train Deep Learning (DL) models by utilizing only the most relevant and informative data for the training process. This approach not only enhances the efficiency of Machine Learning (ML) but also eases the burden of the annotation process [3]. Annotating data for model training can be challenging, time-consuming, error-prone, and expensive. The complexity and cost of this process can escalate significantly, particularly when experts' involvement is necessary for data annotation. Another approach that tackles the challenge of data scarcity is Few-Shot Learning (FSL). This technique aims to enable models to generalize their knowledge and perform similar, yet slightly different tasks using only a few examples for training. This approach is highly relevant in the context of pest detection, where data scarcity is common. It enables us to train new models by leveraging the knowledge of previous models and using only a few available images of the pests for training.

In this research, we adopt a non-Meta-Learning Few-Shot Learning (FSL) approach, specifically Transfer Learning (TL), to investigate its influence on training object detection models. We conduct a comparative analysis of the impact on performance and training time when employing this approach as opposed to not using it.

The paper's structure is organized as follows. The subsequent section offers an overview of the background information relevant to our study. In Sect. 3, we provide a detailed outline of our methodology, encompassing the dataset creation and experiment design. Our findings and the discussion of the results are presented in Sect. 4. Finally, Sect. 5 presents the conclusions drawn from our research and outlines potential directions for future studies.

2 Background

In this section, we lay the groundwork by introducing the essential concepts relevant to our research. We present a study conducted on intelligent pest detection and outline the key principles of object detection and classification models, with a specific emphasis on the one-stage model, YOLOv5. Additionally, we delve into the field of Few-Shot Learning (FSL), with a particular focus on Transfer Learning (TL).

2.1 Pest Detection

Agriculture 4.0, also known as Smart Farming, emerged with the goal of making agriculture more efficient. As the world's population grows, the demand for food has significantly increased. Consequently, the pursuit of more profitable production in this field has become highly relevant. In this research, we focus on pest detection, which is one of the major causes of crop losses in the field of agriculture.

One specific pest that causes significant losses is the whitefly, which prompted the authors of [10] to construct a dataset of images of this pest on yellow sticky traps. In [1], the authors utilized this dataset to build a system for detecting whiteflies on traps using an internal camera connected to the cloud. The system analyzes daily captured images for pest detection. Several object detection models were tested, and YOLOv5 models notably outperformed Faster R-CNN. The authors opted to use YOLOv5 small in the trap due to its speed (0.01 s per image), low memory usage and good performance, achieving an 81.2% mAP.

2.2   Object Detection and Classification

Object detection and classification algorithms are largely based on Convolutional Neural Networks (CNNs). They generally perform two tasks: detecting objects within an image and then classifying them. Object detection models can be broadly categorized into two types: single-stage models and two-stage models.

Single-stage models process the image in a single pass, which often makes these models faster than two-stage models, which analyze the image in two stages. Faster R-CNN [14] and Fast R-CNN [6] are examples of two-stage models, both belonging to the R-CNN algorithm family. In contrast, the YOLO algorithm exemplifies single-stage models. Currently, YOLOv8 is considered the state-of-the-art in object detection algorithms [7]. For the purpose of this research, we are particularly interested in YOLO models due to their high speed and lightweight nature.

YOLOv5 was proposed by Ultralytics and held the state-of-the-art title until its successors were introduced. Figure 1 highlights the three components into which the architecture can be divided [8]: the *backbone*, which extracts the features from the input image; the *neck*, where the extracted features are processed to provide context to next component; and the *head* or *output*, which predicts the locations, classes, and confidences of the objects detected in the image.

The performance of object detection models is typically assessed using mean Average Precision (mAP), offering valuable insights into the model's efficacy in detecting annotated objects. To comprehend mAP, it is essential to understand the concept of Intersection over Union (IoU), which is calculated as follows:

$$Intersection\ over\ Union\ (IoU)\ =\ \frac{A \cap B}{A \cup B}, \tag{1}$$

where A is the real object box and B is the prediction box obtained from the detection model. Basically the numerator represents the $Overlap_{area}$, i.e., the shared region between the predicted and ground truth bounding boxes, and the denominator represents the $Union_{area}$, i.e., the combined area of both these boxes. The predicted bounding box is the object's location predicted by the

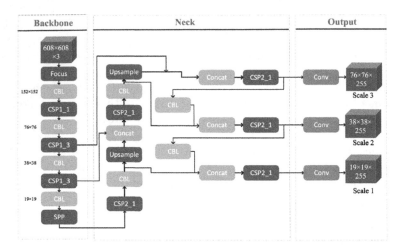

Fig. 1. Overview of YOLOv5 architecture: The model's Backbone extracts relevant features from the input; the Neck processes the extracted features, providing insight to the next component; and the Output is responsible for making the predictions [18].

model and the ground truth corresponds to the manually annotated object location in the image. Using this metric, the model computes True Positives (TPs) and False Positives (FPs). For an IoU threshold of 0.5, objects detected with $IoU >= 0.5$ are classified as TPs, while those with $IoU < 0.5$ are classified as FPs. This allows the calculation of the Precision-Recall (PR) curve across various confidence levels. The mAP is then determined by averaging the Area under the Curve (AuC) of the PR curve across all classes.

2.3 Few-Shot Learning

DL requires a substantial amount of data for training, which makes it an undesirable algorithm in the face of data scarcity. FSL emerged to address the challenge of limited data availability in training DL models, enabling them to learn tasks with just a few annotated examples. This approach originated from the effort to narrow the gap between human-like learning and machine learning [12]. Humans have an innate ability to learn new concepts with little or no previous demonstration. The idea is to systematically revisit previous learning iterations and define a promising strategy based on experience. Specifically, a few examples are needed to make the process faster and more efficient. This inductive-learning perspective gave rise to the few-shot learning methods [4,17]. FSL can be categorized into meta-learning and non-meta-learning algorithms. A meta-learning algorithm aims to learn new representations across few-shot tasks to predict a new set of test tasks with limited available data [5]. A meta-learner iteratively updates model parameters and generalizes to new experimental tasks from a limited amount of labeled data. Specifically, considering a set of training tasks t and task-specific data D_t (meta-training), the goal is to learn model parameters

θ that generalize well across all learning tasks,

$$\theta' = argmin_\theta \sum_{t_i \sim \rho(t)} \mathcal{L}(D_{t_i}, \theta) \tag{2}$$

where $\rho(t)$ is a distribution of tasks, D_{t_i} is the data of task t_i and $\mathcal{L}$ the loss for a downstream task. Model parameters are updated to quickly adapt to the new task at hand. In the meta-testing phase, the meta-learner generalizes to unseen representations by mapping the initialized parameters θ' to new test tasks t_j with parameters

$$\theta'' = \mathcal{L}(D_{t_j}, \theta') \tag{3}$$

where D_{t_j} contains just a few examples of each test task t_j. In general, FSL meta-learning approaches can be broadly divided in two main classes of few-shot models: metric-based and optimization-based methods. Metric-based models adapt to new tasks by learning the similarity between a support set and a query set [15,16]. Optimization-based methods update model parameters across tasks to quickly adapt to new representations with just a few gradient steps [5,13]. The goal is to learn an optimized parameter configuration that generalizes well to new test tasks with just a few examples. In meta-training, the model is trained on the support set of each training task and is evaluated by computing the gradient and loss on a query set. In the meta-testing phase, the updated parameters are used to quickly adapt to new test tasks using limited available data and just a few gradient steps.

In this research, our primary focus is on TL, a non-meta-learning approach. This technique involves leveraging the knowledge that a model has gained from performing tasks that are similar, yet not identical. Figure 2 provides a schematic representation of the TL process in a CNN.

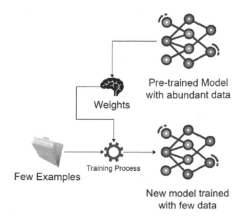

Pre-trained Model
with abundant data

Weights

Few Examples

Training Process

New model trained
with few data

Fig. 2. An Overview of TL in CNNs: A new model is initiated with the weights of a pre-trained model, facilitating its training process

As defined in [11], a domain is denoted as $\mathcal{D} = \{\mathcal{X}, P(X)\}$, where $\mathcal{X}$ represents the feature space and $P(X)$ is the marginal probability distribution. A task $\mathcal{T}$ consists of a label space Y and a predictive function $f(.)$. Given a source domain $\mathcal{D}_s$ and a source task $\mathcal{T}_s$, a target domain $\mathcal{D}_t$ and a target task $\mathcal{T}_t$, transfer learning aims to help improve the learning of the target predictive function $f_t(.)$ in $\mathcal{D}_t$ using the knowledge in $\mathcal{D}_s$ and $\mathcal{T}_s$, where $\mathcal{D}_s \neq \mathcal{D}_t$, or $\mathcal{T}_s \neq \mathcal{T}_t$. In the context model training, TL is commonly implemented by using the weights of a pre-existing model as a starting point. This means that the training process does not start from scratch as the model already possesses some knowledge about performing a similar task. Such an approach conserves time and resources and, in the context of this research, it facilitates the construction of a model with a limited amount of data.

3   Methodology

In this research, we aim to explore and evaluate the effectiveness of transfer learning techniques when applied to object detection models in the field of pest detection. Our primary focus is on whitefly detection as we have the ability to utilize the model from [1] as a starting point to train a new model. This model was trained to detect this pest using high-resolution images (5184×3456 pixels) captured in a controlled environment, i.e., with regulated light exposure and absence of noise, such as yellow sticky traps, as illustrated in Fig. 3.

In this study, we collected 500 images of tomato leaves infested with whiteflies from a greenhouse during the late stages of plant growth. These images, with a resolution of 3000×3000, were taken from three randomly selected rows within the plantation, which was composed of rows of tomato plants. Due to the nature of the greenhouse environment, the images were captured in an uncontrolled setting, resulting in varying light exposures and occasional disturbances, such as out-of-focus instances. Additionally, these images are complex and rich in content, displaying a range of colors and elements, including the green of the leaves, the red of the tomato fruits, the white of the whiteflies, among various other features in the background. Figure 3 provides an illustration of an image from our dataset.

3.1   Experiment Design

To evaluate the effectiveness of TL in the specific case of whitefly detection, we conducted an experiment that involved training several models under different scenarios, varying the number of training images. We start with two images and finish with all the available examples. We planned to train two models with each set of images: one using the weights from the YOLOv5 small model from [1] as a starting point, and the other initializing the weights randomly. We denote the latter as "Scratch" models since their training starts from scratch.

From our dataset, 200 images were annotated and randomly split into 160 images for training and validation, and 40 for testing. For the creation of each

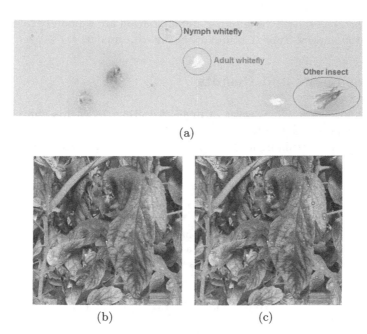

(a)

(b)   (c)

Fig. 3. Comparison of the images from: a) the dataset used in [1], which was taken under regulated light exposure and absence of noise; b) an image collected in our work, which is abundant in content and showcases a variety of colors and elements; and c) the same image as in b) but annotated from our dataset to indicate whiteflies using red squares. In image a), whiteflies are annotated with green circles.

Table 1. Overview of the experiment: The first column indicates the number of images used in the training process, while the subsequent columns represent the distribution of images used for training and validation in each scenario, respectively. Note: The exploration of 1-shot learning, i.e., training with only one example, is not feasible since YOLOv5 requires images for the validation.

Number of images	Training Images	Validation Images
0	–	–
2	1	1
4	2	2
7	5	2
15	10	5
20	15	5
30	20	10
50	40	10
160	120	40

scenario, the images used for training and validation are randomly selected from the set of 160 images. Table 1 summarizes the experiment.

4   Experiments and Results

To account for the random factor in image selection for each scenario, we conducted the experiment five times to ensure the robustness of our findings. Each scenario underwent training for 200 epochs. Table 2 showcases the average mAP and training time obtained over all repetitions for each scenario.

Table 2. Table presenting a comparison of models trained with the same image set but different weight initialization methods: "TL" (Transfer Learning with pre-trained model weights) and "Scratch" (randomly initialized weights). The number of images used, mAP results, and training times are outlined for each approach. Bold figures denote superior performance. The models were trained using an NVIDIA GeForce RTX 4080.

# Images	Average mAP		Average Training Time (s)	
	TL	Scratch	TL	Scratch
0	**0.02**	0.00	–	–
2	**0.16**	0.00	**225**	248
4	**0.18**	0.00	**210**	241
7	**0.19**	0.00	**227**	253
15	**0.35**	0.00	**245**	287
20	**0.42**	0.04	**298**	346
30	**0.56**	0.23	**300**	373
50	**0.62**	0.34	**405**	443
160	**0.77**	0.68	**917**	962

The results clearly illustrate a significant advantage in utilizing TL. Across all scenarios, the application of TL outperforms the random initialization of the CNN weights, in both performance and training time. The results obtained unequivocally demonstrate that the implementation of TL significantly boosted the mAP by an average of 24% and accelerated the training process by 10%.

In Fig. 4, we present a comparative view of the mAP evolution during the training process in one of the scenarios with 160 images. This analysis clearly illustrates the substantial advantage of employing TL. The model trained using TL achieved satisfactory performance as early as epoch 40, while the model trained with randomly initialized weights only reached comparable results at epoch 150. However, it is important to note that no model performed satisfactorily when trained with only a few examples. Exclusively in the final scenario, where the full dataset was used, the model achieved an acceptable performance with an mAP of 0.78. This event might be attributable to the differences in the tasks of each model. The pre-trained model was trained to detect white-flies in a controlled environment with high-resolution images, while our model was trained to detect the same pest but in a completely different environment and with significantly lower resolution. The prior knowledge of the pre-trained model provided a significant advantage in the training process but was definitely insufficient to detect the same pest in an entirely different environment.

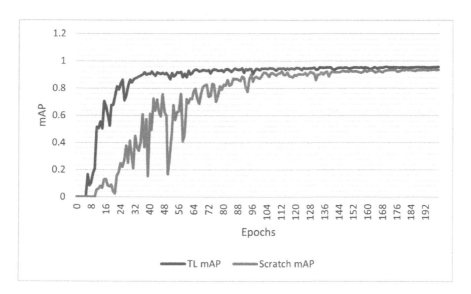

Fig. 4. mAP evolution during training for a scenario with 160 images. The graph depicts two training processes: one using Transfer Learning (TL mAP) and the other with randomly initialized weights (Scratch mAP).

Furthermore, the fact that the models were trained for 200 epochs may not have been enough for them to learn all the features necessary for optimal performance. In contrast, the model was achieving higher mAP values during the training process on the validation set, reaching up to 0.96 mAP, indicative of overfitting. Therefore, finding a suitable balance between the number of epochs, the number of examples, and the distribution of examples in the training and validation sets during the training process is a challenging task. In [2], the same dataset was used to train a model for the same task, and it achieved a significantly higher mAP of 0.931, on the same test set we used. This suggests that selecting the right configuration when training these types of models can indeed be challenging.

To delve deeper into the impact of the number of epochs on the training process, we conducted an additional experiment in a 20-shot scenario (i.e., using 20 images), dividing them into 15 for training and 5 for validation. We varied the number of epochs, starting from 50 and incrementally going up to 700 with intervals of 50. For each epoch scenario, two models were trained using the same set of randomly selected images for the experiment. While one model employed TL, the weights of the other model were randomly initialized. Figure 5 depicts the results obtained from our experiment. Once again, the results demonstrate that models utilizing TL outperform those with randomly initialized weights. However, even with an increase in the number of epochs, the use of TL was not enough build models with satisfactory performance, as the highest mAP achieved was 0.58 with 600 epochs. Despite this, employing TL resulted in an average mAP performance increase of approximately 30% over randomly initialized weights.

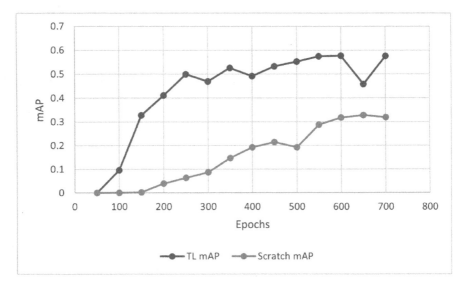

Fig. 5. mAP performance difference between models utilizing Transfer Learning (TL mAP) and those with randomly initialized weights (Scratch mAP), across various numbers of epochs in training.

5   Conclusion

The primary aim of this research was to explore the effectiveness of Transfer Learning (TL) techniques when applied to object detection models. Our findings demonstrated notable advantages in the training process with TL, exhibiting an average increase of 24% in mAP performance and a 10% reduction in training time. However, in cases where only a few examples were used for training, the benefits were not sufficient for successful task completion.

In the realm of Few-Shot Learning (FSL), the Meta-Learning approach seeks to construct models capable of performing tasks when trained with limited examples. By exposing the model to various tasks, it adapts to the learning process, enabling effective learning even with few examples when encountering novel, unseen tasks. Considering that the pre-trained model used in this research was trained solely for whitefly detection, a Meta-Learning approach may be more suitable for the specific case of whitefly detection compared to the TL approach employed here.

Future endeavors should prioritize the exploration of Meta-Learning approaches to identify a strategy that allows for the construction of a robust model with limited training examples, addressing the challenges posed by data scarcity in pest detection scenarios.

Having employed YOLOv5, a lightweight architecture, for our models, we can now leverage it to provide farmers with a reliable tool for whitefly detection. This model can power an IoT camera. When placed in a plantation, this setup helps to monitor and manage the progression of pests, further advancing the initiatives of Agriculture 4.0.

Acknowledgments. This work was supported by project PEGADA 4.0 (PRR-C05-i03-000099), financed by the PPR - Plano de Recuperação e Resiliência and by national funds through FCT, within the scope of the project CISUC (UID/CEC/00326/2020).

References

1. Cardoso, B., Silva, C., Costa, J., Ribeiro, B.: Internet of things meets computer vision to make an intelligent pest monitoring network. Appl. Sci. **12**(18) (2022). https://doi.org/10.3390/app12189397, https://www.mdpi.com/2076-3417/12/18/9397
2. Costa, D., Silva, C., Costa, J., Ribeiro, B.: Enhancing pest detection models through improved annotations. In: Progress in Artificial Intelligence. Springer International Publishing (2023)
3. Costa, D., Silva, C., Costa, J., Ribeiro, B.: Optimizing object detection models via active learning. In: Pertusa, A., Gallego, A.J., Sánchez, J.A., Domingues, I. (eds.) Pattern Recognition and Image Analysis, pp. 82–93. Springer Nature Switzerland, Cham (2023). https://doi.org/10.1007/978-3-031-36616-1_7
4. Fei-Fei, L., Fergus, R., Perona, P.: One-shot learning of object categories. IEEE Trans. Pattern Anal. Mach. Intell. **28**(4), 594–611 (2006). https://doi.org/10.1109/TPAMI.2006.79
5. Finn, C., Abbeel, P., Levine, S.: Model-agnostic meta-learning for fast adaptation of deep networks. In: Precup, D., Teh, Y.W. (eds.) Proceedings of the 34th International Conference on Machine Learning. Proceedings of Machine Learning Research, vol. 70, pp. 1126–1135. PMLR, 06–11 August 2017. https://proceedings.mlr.press/v70/finn17a.html
6. Girshick, R.: Fast r-cnn (2015)
7. Jocher, G., Chaurasia, A., Qiu, J.: YOLO by Ultralytics, January 2023. https://github.com/ultralytics/ultralytics
8. Jocher, G., et al.: Laughing, UnglvKitDe, Sonck, V., tkianai, yxNONG, Skalski, P., Hogan, A., Nair, D., Strobel, M., Jain, M.: ultralytics/yolov5: v7.0 - YOLOv5 SOTA Realtime Instance Segmentation, November 2022. https://doi.org/10.5281/zenodo.7347926, https://doi.org/10.5281/zenodo.7347926

9. Moysiadis, V., Sarigiannidis, P., Vitsas, V., Khelifi, A.: Smart farming in Europe. Comput. Sci. Rev. **39**, 100345 (2021). https://doi.org/10.1016/j.cosrev.2020. 100345, https://www.sciencedirect.com/science/article/pii/S1574013720304457

10. Nieuwenhuizen, A., Hemming, J., Suh, H.: Detection and classification of insects on stick-traps in a tomato crop using faster R-CNN, September 2018, http://nccv18.nl/program/, the Netherlands Conference on Computer Vision, NCCV18; Conference date: 26-09-2018 Through 27-09-2018

11. Pan, S.J., Yang, Q.: A survey on transfer learning. IEEE Trans. Knowl. Data Eng. **22**(10), 1345–1359 (2010). https://doi.org/10.1109/TKDE.2009.191

12. Parnami, A., Lee, M.: Learning from few examples: a summary of approaches to few-shot learning (2022)

13. Ravi, S., Larochelle, H.: Optimization as a model for few-shot learning. In: International Conference on Learning Representations (2017). https://openreview.net/forum?id=rJY0-Kcll

14. Ren, S., He, K., Girshick, R., Sun, J.: Faster R-CNN: towards real-time object detection with region proposal networks (2016)

15. Snell, J., Swersky, K., Zemel, R.: Prototypical networks for few-shot learning. In: Guyon, I., Luxburg, U.V., et al. (eds.) Advances in Neural Information Processing Systems, vol. 30. Curran Associates, Inc. (2017). https://proceedings.neurips.cc/paper_files/paper/2017/file/cb8da6767461f2812ae4290eac7cbc42-Paper.pdf

16. Vinyals, O., Blundell, C., Lillicrap, T., kavukcuoglu, k., Wierstra, D.: Matching networks for one shot learning. In: Lee, D., Sugiyama, M., Luxburg, U., Guyon, I., Garnett, R. (eds.) Advances in Neural Information Processing Systems, vol. 29. Curran Associates, Inc. (2016), https://proceedings.neurips.cc/paper_files/paper/2016/file/90e1357833654983612fb05e3ec9148c-Paper.pdf

17. Wang, Y., Yao, Q., Kwok, J.T., Ni, L.M.: Generalizing from a few examples: a survey on few-shot learning. ACM Comput. Surv. **53**(3) (2020). https://doi.org/10.1145/3386252

18. Zhu, L., Geng, X., Li, Z., Liu, C.: Improving yolov5 with attention mechanism for detecting boulders from planetary images. Remote Sens. **13**(18) (2021). https://doi.org/10.3390/rs13183776

Do Emotional States Influence Physiological Pain Responses?

Bruna Alves[1,2](✉) ⓘ, Catarina Silva[2], and Raquel Sebastião[1,3] ⓘ

[1] IEETA, DETI, LASI, University of Aveiro, 3810-193 Aveiro, Portugal
[2] Department of Physics (DFis), University of Aveiro, 3810-193 Aveiro, Portugal
[3] Polytechnic of Viseu, 3504-510 Viseu, Portugal
{bruna.alves,catarinavilar,raquel.sebastiao}@ua.pt

Abstract. Pain is a highly subjective and complex phenomenon. Current methods used to measure pain mostly rely on the patient's description, which may not always be possible. This way, pain recognition systems based on body language and physiological signals have emerged. As the emotional state of a person can also influence the way pain is perceived, in this work, a protocol for pain induction with previous emotional elicitation was conducted. Eletrocardiogram (ECG), Electrodermal Activity (EDA) and Eletromyogram (EMG) signals were collected during the protocol. Besides the physiological responses, perception was also assessed through reported-scores (using a numeric scale) and times for pain tolerance. In this protocol, 3 different emotional elicitation sessions, negative, positive and neutral, were performed through videos of excerpts of terror, comedy and documentary movies, respectively, and pain was induced using the Cold Pressor Task (CPT). A total of 56 participants performed the study (with 54 completing all three sessions). The results showed that during the negative emotional state, pain reported-scores were higher and pain threshold and tolerance times were smaller when compared with positive. As expected, the physiological response to pain remain similar despite the emotional elicitation.

Keywords: Cold Pressor Task · Emotion · Pain · Physiological Signals

1 Introduction

Pain is a subjective phenomenon that depends on the past experiences of each individual and the circumstances of the moment. It's a survival mechanism that allows us to identify harmful situations and avoid tissue damage [9].

Pain has a big impact on people's lives and society in general, it's the principal reason for seeking medical attention and it can also provoke a loss of productivity in companies [9]. Moreover, chronic pain costs society more than cancer and heart diseases [9]. So, it is important to deal with pain as soon as possible, identifying its origin to achieve diagnosis and adequate treatment, preventing harmful consequences.

ⓒ Springer Nature Switzerland AG 2024
V. Vasconcelos et al. (Eds.): CIARP 2023, LNCS 14470, pp. 117–131, 2024.
https://doi.org/10.1007/978-3-031-49249-5_9

Currently, there are several methods to measure pain, but all of them depend on the patient's description. Pain assessment is typically done by a caregiver through self-reports, observing behavioral or physiological pain responses, and using information about the pain cause [9]. The methods used to quantify pain are usually visual or numeric scales [4]. However, patients with limited communication skills cannot report their pain experience, these may include infants and children, adults with cognitive damage or intellectual disability, and unconscious people [9]. Thus, an objective measurement of pain could be beneficial. To achieve this goal, there has been some research devoted to the development of pain recognition systems, which are based on the detection of some characteristics provoked in the human body by pain, such as facial expressions, sounds, gestures, or even some physiological signals.

To evaluate pain sensitivity there are three measurements that are commonly used: pain threshold, which is the minimal stimulus intensity required to elicit pain; pain tolerance, which is the maximal stimulus intensity that an individual can withstand; and pain perception, which refers to what classification an individual gives to a standardized stimulus intensity [10]. Pain is not only a physical experience but is also connected with emotions. As mentioned earlier, pain is a subjective experience that depends on the circumstances in which it occurs. One factor that can influence the pain experience is the emotional state of the individual [4].

This work addresses the influence of emotional states on pain responses. Firstly, the participants are subjected to the elicitation of different kinds of emotions, namely negative, positive, and neutral, through the visualization of different excerpts of terror, comedy and documentary movies, respectively, while pain induction is attained through the Cold Pressor Task (CPT) test in three different emotional sessions. Throughout each entire session, electrocardiogram (ECG), eletromyogram (EMG) from triceps and trapezius, electrodermal activity (EDA) and pain-reported measures are collected.

The aim of this work is to understand if emotion elicitation has an influence in pain perception and response. It is expected that emotional elicitation will not have an influence on the physiological response to pain, while it is supposed that perception depends on the elicited emotional state. Negative emotions (in this case, fear) should exacerbated pain, increasing the perception and lowering the tolerance to pain. On the other hand, positive emotions (happiness) should attenuate the perception of pain. This way, this work proposes a new protocol to assess pain perception related to the emotional state of a person and establishes a relationship among pain and positive and negative emotions.

This document is structured as follows: Sect. 2 presents some related studies that assess emotional modulation in pain perception; Sect. 3 describes the materials and the methods used to develop this work; Sect. 4 exposes the obtained results; Sect. 5 presents its discussion and Sect. 6 highlights the main conclusions achieved and a perspective for future research.

2 Related Works

There have been several studies developing pain recognition systems with different approaches. The works below support that emotions can have a crucial role in pain perception, driving the motivation for the present work in studying the influence of emotions on pain perception.

Zhang *et. al* [10] investigated the differences in pain perception between men and women and how that is related to negative emotions. To quantify pain sensitivity, they used a cold pressure test (CPT), and to assess negative emotions they used several questionnaires and MRI data. The hypothesis was that females experience more negative emotions and that is related to a higher pain sensitivity. So, first, the subjects responded to emotion-related questionnaires and then they were submitted to a CPT. The questionnaires included were the Chinese version of the Fear of Pain Questionnaire (FPQ), the Pain Anxiety Symptoms Scale-20 (PASS), the trait version of State-Trait Anxiety Inventory (STAI) and the Beck Depression Inventory. The pain was induced by putting the individual's left hand into cold water at a temperature of $2°C$. Pain threshold was determined as the duration of immersion from the time that the hand was kept in the water to the time the subject began to feel pain and pain tolerance as the total time from immersing the hand in the water to the time the participants remove it. Several statistical analyses were performed, such as the non-parametric Mann-Whitney U-tests and Spearman rank correlation analyses (with the skewed distributed variables) and the parametric independent samples t-tests and Pearson correlation analyses (with normally distributed variables). They found statistically significant gender differences in pain threshold and tolerance, and two questionnaires: the males presented higher pain threshold and tolerance and lower scores in the FPQ and PASS questionnaires. Further analysis showed that the two questionnaires' scores were negatively correlated with pain threshold and tolerance, showing that differences in pain sensitivity were mediated by pain-related negative emotions, specifically pain-related fear and anxiety.

Silva and Sebastião [7] studied the ECG signal during pain induction under disctint emotional contexts. The participants were subjected to a CPT, for inducing pain, while watching an emotional inducing video. The protocol consisted of two sessions, one using a fear emotion-inducing video and a second using a neutral one. The data was pre-processed and then was used 8 machine learning algorithms to classify pain. Attempting on a binary classification of pain, training and testing was performed in several strategies using ECG data from both emotional sessions, and classification results were compared across different strategies. In their work, the results supports that ECG response remain similar along both sessions. As for the classifiers, the RF and AdaBoost showed better performance to classify pain and the LDA and LR models were the worst ones.

Srisopa *et. al* [8] found that emotion regulation strategies produced significant improvements in decreasing pain intensity during labor. In this review, the type of pain studied was the pain caused in the labor and was measured by self-report or the observation of the participant's behavior. Strategies based on mindfulness intervention and distraction were used to train the subjects to manage pain. The individuals submitted to these techniques showed a significantly reduced pain intensity during the active phase of labor.

Although the studies of Zhang *et al.* [10] and Srisopa *et al.* [8] associate emotional state with pain perception, they do not establish a protocol for emotional and pain induction. The protocol proposed by Silva *et al.* [7] has certain limitations, which will be discussed in Sect. 5.

3   Materials and Methods

This section presents the materials and methods implemented in this work. The data acquisition and approaches used to analyse the data are also exposed.

3.1   Experimental Setup

Pain was induced using the Cold Pressor Task (CPT). The CPT is a test that involves putting a hand or forearm in cold water causing a stimulus that produces a slowly increasing pain of slight to moderate intensity. It has been used in many different types of works, such as studies about pain, autonomic reactivity, and hormonal stress responses [2].

For the signals collection, a 4-Channel Biosignalsplux[1] was used. Four sensors were connected to this device: two EMG sensors, one EDA sensor, and one ECG.

In Fig. 1 the equipment's setup and its placement in the room are shown.

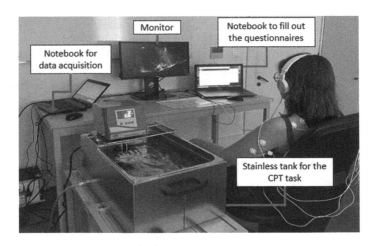

Fig. 1. Experimental Setup.

For the CPT test, a stainless-steel tank of 45 liters was used. The tank has an immersion thermostat, that includes a circulation pump that can be used to

---

[1] https://www.pluxbiosignals.com/apps/builder/biosignalsplux-kit-builder (accessed 20 July 2023).

improve the homogeneity within the bath and to perform a closed liquid circulation circuit. The water was initially cooled with ice and the control panel allows the control and adjustment of temperature. Also, two notebooks were used: one for the acquisition of the sensors' reading and the respective monitorization (through the software OpenSignals from Biosignalsplux) and the other for displaying the videos with an external monitor (used with headphones for audio). This notebook was also used for completing the questionnaires. The experimental procedure was implemented at the Institute of Electronics and Informatics Engineering of Aveiro, University of Aveiro, in a room prepared specifically for this purpose.

3.2   Protocol

Before the procedure, informed consent with all the information about the process was given to the participants. In the case of a positive response, the first questionnaire, the trait version of State-Trait Inventory for Cognitive and Somatic Anxiety (STICSA-trait) was also given to the participant to be replied before the procedure.

The protocol begins with the participant answering the following questionnaires: the state version of STICSA (STICSA-state), the Perceived Stress Scale (PSS), the Eysenck Personality Inventory (EPI) and the Visual Analog Scale (VAS), in this order. The purpose of these questionnaires is to assess some psychological traits of the participant, as well as the emotional state. The VAS questionnaire measures the participant's arousal and valence state.

After the questionnaires, the electrodes connected to the Biosignalsplux sensors are placed on the participant according to Fig. 2.

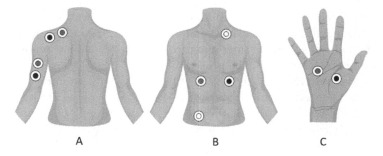

Fig. 2. Scheme of the electrodes placement: A) EMG electrodes on the trapezius and triceps muscles; B) ECG electrodes plus reference electrode of the EMG and C) EDA electrodes.

The positive electrodes are represented with a red circle and the negative ones with a black circle. The white electrodes are the reference electrodes for the ECG and EMG signals and must be placed above the pelvic bone for ECG above on the collarbone for EMG. The EMG electrodes were placed in the trapezius and triceps muscles of the non-dominant arm. The ECG electrodes were placed on the rib cage (the positive one on the right side and the negative one on the left side of the body). The EDA was collected on the dominant hand: the positive electrode was placed in the upper part of the palm and the negative in the lower part of the palm.

Accounting for the inter-participant variability regarding the physiological responses, the data collection begins with a rest time (Baseline 1), corresponding to five minutes, where the person will be just sat in a comfortable position without any stimuli. After this time, while still in a comfortable position without any pain stimuli, the participant will watch an emotional-inducing video for around ten minutes, in the frontal screen. The video can be one of three kind: an neutral emotional inducing video, which is composed of excerpts of documentaries, a negative emotional inducing video (Fear), which is composed of excerpts of terror movies or a positive emotional inducing video (Happiness), which is composed by excerpts of comedy movies.

When the video ends, the subject will be asked to respond to a last questionnaire, the VAS-pos, and then another rest time begins (Baseline 2) also with a duration of five minutes. At the end of this resting time, the participant is asked to report the pain level, in his non-dominant hand, using a numerical pain scale (NPS) ranging from 0 to 10. Afterwards, the pain stimulus is applied. The individual will be requested to put his non-dominant hand in the cold-water tank with a temperature of approximately $7°C \pm 1°C$, beginning the CPT test. To register the participant's pain threshold, they are asked to report the pain level using the NPS as soon as they feel any pain. The participants are informed to hold the hand immersed as long as they can, with a time limit of 2 min. If they reach the point where they can no longer tolerate the pain, they are instructed to report to the researchers that they will remove the hand from the tank. Before doing so, they are asked to report the pain level (pain tolerance). If they can keep the hand immersed in the tank for the complete duration, the maximum pain experience will be reported at the limited time defined (2 min). Finally, the last rest period begins. After three minutes the level of pain is reported again. At the end of the rest period, the procedure ends.

Each participant repeats this protocol three times, with an interval of approximately 1 week, where each session differs from the type of emotion-induced through the video. The order of the videos is randomized. In the second and third sessions, the participants will only respond to the questionnaires STICSA-state, VAS-pre, and VAS-pos.

The protocol for data collection is schematized in Fig. 3.

This study was approved by the Ethics and Deontological Council of the University of Aveiro (CED-UA-12-CED/2023).

3.3 Physiological Data Preparation

After the acquisition, the physiological data was filtered and divided into epochs according to the triggers given and processed.

Fig. 3. Scheme of the protocol applied.

The epochs used in this work correspond to Baseline 1 and CPT epochs only, since the aim is to evaluate if the emotion elicitation had some influence on physiological response to pain and in pain perception.

The signals were pre-processed using Neurokit2[2] in Python. The ECG was filtered using a 5th-order high-pass Butterworth filter with cut off frequency of 0.5 Hz, followed by powerline filtering (50 Hz). The EDA was filtered using a 4th-order low-pass Butterworth filter with cut off frequency of 5 Hz followed by smoothing of the signal. Lastly, the EMG was filtered using a 4th-order 10 Hz highpass Butterworth filter followed by a notch filter at 50 Hz and a constant detrending. After the filtering, Neurokit2 functions were used to extract some important features of the signals.

Table 1 summarizes all the features extracted from the physiological signals. The choice of the features was based on previous studies where CPT was used [6,7] and other features were considered for analyses.

Regarding the Heart Rate Variability (HRV), only ultra-short metrics were extracted since the CPT lasted 2 min or less. The works of Salahuddin *et. al* [5] and Boonnithi *et. al* [3] prove that the HRV features present in Table 1 are suitable to be calculated through signals lasting only 30 s or less. This way, only the sessions where the participant endured at least 30 s with the hand on the cold water tank were tacked into account to this work.

In order to minimize inter-participant variability, the features were normalized by the ratio between those features extracted from CPT and those extracted from Baseline 1, for each participant.

As HRV_pNN50 had some zero values in the Baseline 1 epoch, this feature was removed from the dataset, since the ratio would lead to NaN values. The HRV_LF, HRV_LFn and HRV_LFHF features were also removed since they had several NaN values.

3.4   Statistical Analysis

In order to investigate if the extracted features differed significantly regarding the emotion elicitation, statistical tests were performed.

First, the normality of all the features was tested using the Shapiro-Wilk test, which tests the null hypothesis that the data was drawn from a normal distribution. Therefore, if the p-value is below a chosen significance level (in this

---

[2] https://neuropsychology.github.io/NeuroKit/ (Accessed 9 July 2023).

Table 1. Description of the extracted features.

Signal	Series	Designation	Description
ECG	HR	Mean_HR	Mean of Heart Rate (HR)
	HRV	RMSSD	Square Root of the Mean of the Squared Successive Differences between adjacent RR intervals
		meanNN	Mean of the RR intervals
		SDNN	Standard Deviation of the RR intervals
		SDSD	Standard Deviation of the Successive Differences between RR intervals
		CVNN	Standard deviation of the RR intervals divided by the mean of the RR intervals
		pNN50	Proportion of RR intervals greater than 50ms, out of the total number of RR intervals
		TINN	Baseline width of the RR intervals distribution obtained by Triangular Interpolation, where the error of the least squares determines the triangle
		HTI	HRV Triangular Index, measuring the total number of RR intervals divided by the height of the RR intervals histogram
		LF	Spectral power of Low Frequencies
		HF	Spectral power of Low Frequencies
		LFHF	Ratio obtained by dividing the Low Frequency power by the High Frequency power
		LFn	Normalized Low Frequency, obtained by dividing the low frequency power by the total power
		HFn	Normalized High Frequency, obtained by dividing the high frequency power by the total power
		SD1	Standard Deviation perpendicular to the line of identity. It is an index of short-term RR interval fluctuations
		SD2	Standard Deviation along the identity line. I ndex of long-term HRV changes
		ApEn	Approximate Entropy
		SampEn	Sample Entropy
	Peaks (P_, T_, R_, S_)	NPeaks/min	Number of peaks per minute
		Amp	Amplitude of the correspondent peak
		dist	Distance (in samples) between consecutive peaks
	Waves (P_, T_, R_)	OnsetAmp	Amplitude of the correspondent waves' onsets
		OffsetAmp	Amplitude of the correspondent waves' offsets
		OnOffDist	Distance (in samples) between consecutive waves onsets and waves' offsets
EDA	SCR	NPeaks/min	Number of peaks per minute
		Mean_SRC	Mean SCR
		SCR_Height	Mean SCR height
		SCR_Amp	Mean SCR amplitude
		SCR_RiseTime	Mean Rise Time
		SCR_RecoveryTime	Mean Recovery Time
	SCL	Mean_SCL	Mean SCL
EMG (Trap_, Tric_)	EMG	Var	Variance of the EMG signal
		RMSE	Root Mean Square for EMG
	Amplitude	Mean_Amp	Mean of the EMG Envelope
		Med_Amp	Median of the EMG Envelope
		RMSA	Root Mean Square for EMG Envelop

case, $\alpha=0.05$), the null hypothesis should be rejected and therefore the feature is not likely to follow a normal distribution [1].

The features that fail the Shapiro-Wilk test (the data does not meet the assumption of normality) were submitted to a non-parametric Friedman test. Those who passed the Shapiro-Wilk test (the data is likely to follow a normal distribution) were submitted to the parametric repeated measures ANOVA test. Both Friedman and ANOVA tests were used to evaluate if the features could differentiate between sessions with different emotion elicitation (F: Fear, H: Happiness, N: Neutral). Afterwards, the Nemenyi post-hoc test was performed to evaluate, for those features, which pair of emotional states differed.

4   Results

A total of 56 volunteers (28 females) with ages between 18 and 30 y.o. (mean of 22.46 and standard deviation of 2.04 y.o.) participate in the study. Only 2 participants did not undergo the last session due to personal reasons. Therefore, a total of 166 sessions were performed.

4.1   Pain Perception

Among all the sessions the mean time of the CPT was about $91.70 \pm 39.64\,s$ (mean±standard deviation). Within the 166 sessions, there were 9 sessions where the participant kept the hand in the cold-water tank for less than 30 s. On the opposite, there were 96 sessions where the participants endured the maximum times (2 min) with the hand immersed.

With regard to the pain scores reported, none of the participants felt pain before the CPT. At the time the participants felt pain (pain threshold) they reported a score of 4.78 ± 2.08 in the NPS. At the end of the 2 min or at the moment the participant took the hand off the tank (tolerance) the pain reported was of 7.84 ± 1.71 in the NPS. After 3 min of the removal of the hand from the cold water, the pain reported decreased to 0.72 ± 1.09 in the NPS. There were 96 sessions where participants did not feel any pain at this time.

Taking into account the emotional elicitation, only data from 54 participants were analysed, since 2 did not perform the three sessions and therefore, they were missing an elicitation.

Considering the negative state, the scores reported at pain tolerance and at the 3 min after taking the hand off the water, were greater than those reported when in neutral and positive states. With respect to the scores reported at pain threshold, the values are similar across emotional sessions, specially for negative and positive inducing sessions (4.70 ± 2.20 and 4.71 ± 2.05, respectively). However, with regard to time, both pain threshold and tolerance were lower for the negative state (15.76 ± 9.29 and 92.11 ± 39.76, respectively) when compared with the positive induced condition (19.74 ± 21.94 and 93.23 ± 38.71, respectively).

Figure 4 presents the violinplots for the reported pain scores (top) and for the times (in seconds) of pain threshold and tolerance (bottom). Regarding the scores, when comparing the violins for each emotion, it can be noted that they are very similar to each other. Only smaller differences, regarding gender, can be found, as females tend to report highest scores than males.

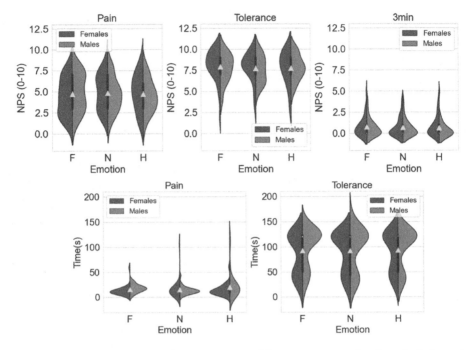

Fig. 4. Violinplots of the NPS scores reported by participants (top) and of the pain threshold and tolerance, expressed in seconds (bottom). The ▲ stands for the mean of the scores and times within each emotion. (Color figure online)

With respect to tolerance's time, the violins are quite similar. However, it is interesting that the three plots seem to have two clusters. It shows that the distribution of the time that participants can stand with the hand immersed in the water is not uniform, indicating that are mainly two groups of participants: those who can tolerate the 2 min and those who can barely reach the 1 min. However, it is evident that more participants can reach the two minutes than the opposite. Regarding gender, no differences stand out, since the violins look quite symmetric.

Concerning pain threshold's time, the violins are also very similar but the values are more dispersed for Positive and Neutral states than for Negative, and, in general, females tend to report pain sooner than males.

These results seem to support the hypothesis that emotion elicitation influences pain perception since, despite the similar scores, when in the negative induced session the pain scores were higher and the threshold and tolerance times where lower, which indicates less resilience to pain and greater exacerbation.

4.2 Physiological Data

As mentioned above, in this work only the sessions where the participant lasted at least 30 s with the hand immersed in the cold water were considered for physiological data analysis. Therefore, the 9 sessions where participants did not endure the CPT for at least 30 s were removed. Two cases where the acquisition of physiological data failed were also removed, leaving signals of 155 sessions for analysis. For statistical analyses, only the participants who had the features computed for the three emotions were considered, leaving a total of 46 participants and therefore 138 samples per feature.

Only two features could differentiate statistically between emotional states, namely R_OnOffDist and SCR_Height.

Figure 5 presents the boxplots of R_OnOffDist and SCR_Height and the p-values between the different emotional states.

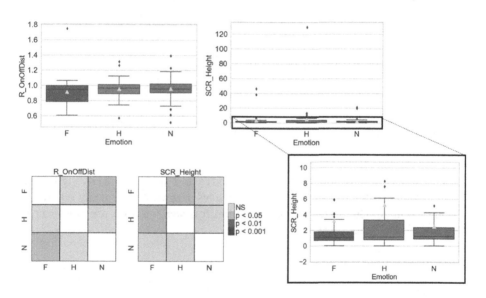

Fig. 5. Boxplots of the features R_OnOffDist and SCR_Height (top) and the respective p-values between different emotions (bottom). The ♦ stands for outliers and the ▲ stands for the mean of the features within each emotion. (Color figure online)

The boxplots of the ECG feature for Positive and Neutral states are quite similar, with the Neutral one presenting slight higher values. However, the boxplot for Negative indicates that the values obtained with this state were lower.

In fact, all the quartiles, and even the mean, are lower in this case. This feature shows a significant statistically difference between the Negative and Neutral states with a p-value lower than 0.05. With regard to SCR_Height (obtained through EDA), this feature differentiated between Fear and Happy emotions. In fact, despite the minimum and median values of the three boxes being very closer, the boxplot of the Happy elicitation presents higher values compared to the other boxplots, while the Fear elicitation presents the lowest values. The means of the three states are highly influenced by the outliers. Furthermore, the two features, specially SCR_Height, have a considerable number of outliers which can explain the unexpected statistical differences.

Figure 6 presents the boxplots of several features obtained from the ECG signal. None of these features differentiates between any emotion elicitation. Despite some slight differences, the boxplots and the means of these features are very similar across the different emotional states.

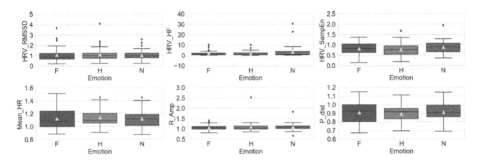

Fig. 6. Boxplots of several features of the ECG signal. The ◆ stands for outliers and the ▲ stands for the mean of the features within each emotion. (Color figure online)

Fig. 7 presents boxplots from features extracted from the EDA signal, and none presents statistical significant differences regarding emotion elicitation. Although some outliers, the means computed for each emotion are close to each other and distributions are similar.

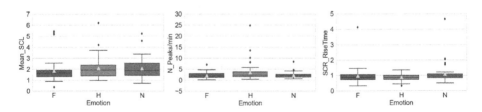

Fig. 7. Boxplots of several features of the EDA signal. The ◆ stands for outliers and the ▲ stands for the mean of the features within each emotion. (Color figure online)

Figure 8 shows the boxplots of features computed from the EMG signals from both the trapezius and triceps muscles, presenting a great number of outliers. For the same feature, the values obtained from triceps muscle are lower than those obtained from trapezius, which indicates that trapezius was more activated. Regarding the features from the triceps muscle, despite, consistently, presenting lowest values for Neutral state, the Friedman test did not find any statistical difference between emotional conditions. For the features obtained through the EMG from trapezius, the values and the boxplots are similar for the three emotional elicitation, and Friedman test did not report significant differences between emotional states for any feature.

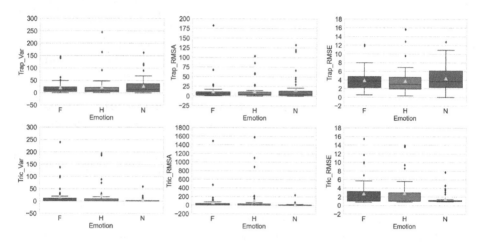

Fig. 8. Boxplots of several features of the EMG signal of both triceps and trapezius muscles. The ♦ stands for outliers and the ▲ stands for the mean of the features within each emotion.

The obtained results for the four physiological signals analyzed support the hypothesis that the physiological response to pain is not influenced by emotional elicitation.

5   Discussion

The aim of this work was to show that emotional elicitation influences the way pain is perceived, but it has no influence on physiological response to pain.

The related works presented supports the idea that emotion has a correlation with pain perception, namely the works of Zhang *et. al* [10] and Srisopa *et. al* [8]. Both articles studied the influence of the emotional state on pain perceived (with different pain origins) and the results lead to the conclusion that negative emotions affect the way pain is perceived [10] and emotion modulation can help to manage pain [8].

In the work of Silva and Sebastião [7] the emotion elicitation was performed during the CPT task. The authors found that this strategy was not adequate since many of the participants did not pay enough attention to the video due to the pain felt at the same time. The current protocol was implemented considering the strategy of watching the video before the CPT, anticipated that this adjustment would elicit emotions. However, although differences regarding scores and times for pain tolerance and threshold, the discrepancy in pain perception was not highlighted.

Also, despite the protocol being thoroughly explained to the participants and all the doubts being addressed, some participants did not understand that they were required to report the time they first perceived pain. Consequently, they only reported pain when specifically asked to do so. This may lead to increased values of pain perception in some cases. By re-evaluating the violinplots of pain perception (Fig. 4), this observation is evident from the long tails of the violins. Additionally, some participants reported being confused when evaluating the pain felt using the NPS, struggling to choose the appropriate number.

Another factor that may have hindered emotional elicitation is the fact that some participants had already watched the movies from which the excerpts were taken. Consequently, in this case, the video was unable to elicit emotion, especially fear.

With regard to gender, the distribution of the violinplots for scores and pain threshold seems to be in line with the conclusions of Zhang *et. al* [10] that males present higher pain threshold and tolerance.

Concerning the second hypothesis, there were only two features among the 52 studied that could distinguish between elicited emotions. This supports the hypothesis that the emotion elicited does not influence on the physiological response to pain. The physiological system shows the same response to pain regardless the emotional state.

6   Conclusions and Further Research

In this work, a protocol for pain induction with previous emotional elicitation was conducted.

The state of the art shows that the emotional state of a person influences the way pain is perceived but has no influence on physiological pain response.

The results showed that emotion elicitation was not clearly achieved, since the pain perception was slightly modified despite the emotional video visualized, which may be due to the time elapsed between the emotional elicitation and the pain induction.

On other hand, as expected, the results showed emotional states pose no influence on the physiological response to pain.

Regarding the encouraging results, further research should be concerned with the design of a protocol to specifically attain the emotional elicitation in order to ensure that emotional states still elicited during pain induction.

With respect to the physiological response to pain, the collected data should be deeper analyzed in order to find relevant patterns and to extract important information for pain prediction. It would also be worthwhile to obtain feature validation from clinical experts and conduct a selection process based on these inputs. Moreover, participant-independent strategies for training and testing the models should be considered, as well as attaining the development of personalized models fitted only with data from the same participant.

As this work emerges within the scope of the EMPA project, the database, fully anonymized, will become available to the scientific community once data collection is complete.

Acknowledgements. This work was funded by national funds through FCT - Fundação para a Ciência e a Tecnologia, I.P., under the grant UI/BI62/10827/2023 (B.A.), and the Scientific Employment Stimulus CEECIND/03986/2018 (R.S.) and CEECINST/00013/2021 (R.S.). This work is also supported by the FCT through national funds, within the R&D unit IEETA/UA (UIDB/00127/2020) and under the project EMPA (2022.05005.PTDC). We extend our sincere appreciation to all the volunteers who participated in the study, as well as to Ana Carolina Almeida and Daniela Pais for their collaboration in data collection and pilot tests.

References

1. Scipy.stats.shapiro - Scipy v1.11.1 manual. https://docs.scipy.org/doc/scipy/reference/generated/scipy.stats.shapiro.html, Accessed 10 Jul 2023
2. von Baeyer, C.L., Piira, T., Chambers, C.T., Trapanotto, M., Zeltzer, L.K.: Guidelines for the cold pressor task as an experimental pain stimulus for use with children. J. Pain **6**(4), 218–227 (2005)
3. Boonnithi, S., Phongsuphap, S.: Comparison of heart rate variability measures for mental stress detection. In: 2011 Computing in Cardiology, vol. 38, pp. 85–88 (2011)
4. Lumley, M.A., et al.: Pain and emotion: a biopsychosocial review of recent research. J. Clin. Psychol. **67**(9), 942–968 (2011)
5. Salahuddin, L., Cho, J., Jeong, M.G., Kim, D.: Ultra short term analysis of heart rate variability for monitoring mental stress in mobile settings. In: 29th Annual International Conference of the IEEE Engineering in Medicine and Biology Society, IEEE, August 2007. https://doi.org/10.1109/iembs.2007.4353378
6. Sebastião, R., Bento, A., Brás, S.: Analysis of physiological responses during pain induction. Sensors **22**(23) (2022). https://doi.org/10.3390/s22239276
7. Silva, P., Sebastião, R.: Using the electrocardiogram for pain classification under emotional contexts. Sensors **23**(3) (2023). https://doi.org/10.3390/s23031443
8. Srisopa, P., Cong, X., Russell, B., Lucas, R.: The role of emotion regulation in pain management among women from labor to three months postpartum: an integrative review. Pain Manag. Nurs. **22**(6), 783–790 (2021)
9. Werner, P., Lopez-Martinez, D., Walter, S., Al-Hamadi, A., Gruss, S., Picard, R.W.: Automatic recognition methods supporting pain assessment: A survey. IEEE Trans. Affect. Comput. **13**(1), 530–552 (2022)
10. Zhang, H., Bi, Y., Hou, X., Lu, X., Tu, Y., Hu, L.: The role of negative emotions in sex differences in pain sensitivity. Neuroimage **245**(118685), 118685 (2021)

Author Index

Printed in the United States
by Baker & Taylor Publisher Services